Heavy Metal
Contamination of Soil
Problems and Remedies

Heavy Metal Contamination of Soil
Problems and Remedies

Editors

Iqbal Ahmad
Department of Agricultural Microbiology
Faculty of Agricultural Science
Aligarh Muslim University
Aligarh
INDIA

S. Hayat
Plant Physiology Section
Department of Botany
Aligarh Muslim University
Aligarh
INDIA

John Pichtel
Ball State University
Natural Resources and Environmental Management
Muncie, IN
USA

Science Publishers, Inc.
Enfield (NH), USA   Plymouth, UK

SCIENCE PUBLISHERS, INC.
Post Office Box 699
Enfield, New Hampshire 03748
United States of America

Internet site: *http://www.scipub.net*

sales@scipub.net (marketing department)
editor@scipub.net (editorial department)
info@scipub.net (for all other enquiries)

ISBN 1-57808-385-0

© 2005, Copyright Reserved

Published by Science Publishers, Inc., Enfield, NH, USA
Printed in India.

Preface

Increased industrialization and human activities have impacted the environment through the disposal of wastes containing heavy metals. Mine drainage, metal processing industries, domestic effluents, petroleum refining, dye and leather industries, agricultural runoff and other sources contribute a wide range of metallic wastes to the biosphere. Heavy metal releases to soil have been increasing continuously and posing significant threats to the environment and public health because of their toxicity and accumulation in soil and in food chains. High levels of heavy metals have been detected in soil due to long-term application of metal-enriched sewage sludge, poorly treated wastewater and industrial effluents, improper disposal of solid waste, atmospheric deposition, and application of agrochemicals in modern agricultural practice.

Once soil is contaminated with heavy metals, it is a difficult and costly affair to remove them. Despite the institution of guidelines and regulatory controls for wastewater and sewage sludge application, there is still only limited knowledge of the adverse effects of long-term application of such wastes on soil health, soil quality and crop productivity. This is partly due to the lack of field data on levels of metal pollution agricultural land and the complex interaction of heavy metals with soil constituents. However, current research can provide insights about potential risk and provide a sound basis for framing regulations and codes of practice for metal-containing wastewaters, which will minimize potentially detrimental effects on the soil-ecosystem and human health.

Scientific research involves a multidisciplinary approach to understanding soil metal accumulation, interaction with soil constituents, metal-microbe interactions and behavior and toxicity of metals in soil-plant systems and for developing effective technologies for remediation. The main purpose of this book is to provide a comprehensive source of current literature about the impact of heavy metal pollution of agricultural soils primarily resulting from long-term application of wastewater, industrial effluents and sewage sludge, and atmospheric deposition. Soil health, soil-microbe interactions, heavy metal accumulation in soil, behavior of metals in soil and remediation will be addressed.

This book is not intended to be an encyclopedic review of the subject matter. However, various chapters incorporate both theoretical and practical aspects and will provide baseline information for future research through which significant developments are possible. It is intended that this book will be useful to students, teachers and researchers, both in universities and research institutes, especially in relation to the environmental and soil sciences. With great pleasure, we extend our sincere thanks to all the contributors for their timely response, their excellent and up-to-date contributions and consistent support and cooperation. We express deep gratitude to Professor Aqil Ahmad, Dr. Zaki A. Siddiqui, Dr. Q. Fariduddin and Mr. Barket Ali, who have been a great source of inspiration. We express our sincere thanks to our wives Mrs. Seema Iqbal, Mrs. Ayesha Hayat and Mrs. Theresa Pichtel, respectively, for all the support they provided and the neglect they suffered during the preparation of this book. The financial assistance rendered by Department of Science and Technology, Government of India, New Delhi to S. Hayat is greatly acknowledged. Finally, we are acknowledged to Almighty God who provided all the channels to work in cohesion and coordination right from the conception of the idea to the development of the final version of this treatise, "Heavy Metal Contamination in Soil: Problems and Remedies".

Iqbal Ahmad
S. Hayat
John Pichtel

Contents

Preface   *v*

Contributors   *ix*

1. Practical Issues of Land Application of Biosolids   1
   P.K. Wong

2. Bioavailability of Metals and Metalloids in
   Terrestrial Environments   25
   Patrick K. Jjemba

3. Microbially Mediated Changes in the Mobility of
   Contaminant Metals in Soils and Sediments   43
   Flynn Picardal and D. Craig Cooper

4. Influence of Long-Term Application of Treated
   Oil Refinery Effluent on Soil Health   89
   Iqbal Ahmad, S. Hayat, A. Ahmad, A. Inam and Samiullah

5. Heavy Metals in Temperate Forest Soils: Speciation,
   Mobility and Risk Assessment   105
   G. Koptsik, S. Lofts, E. Karavanova, N. Naumova and M. Rutgers

6. Bioremediation   157
   Andrew S. Ball

7. Effects of Fly Ash on Soil Characteristics, Plant
   Growth and Soil Microbial Populations   171
   Zaki A. Siddiqui and Lamabam P. Singh

8. Effects of Metal-Contaminated Organic Wastes on
   Microbial Biomass and Activities: A Review   195
   K. Chakrabarti, P. Bhattacharyya and A. Chakraborty

9. Characterization and Evaluation of Municipal Solid Waste
   Compost by Microbiological and Biochemical Parameters
   in Soil under Laboratory and Field Conditions   205
   P. Bhattacharyya, K. Chakrabarti and A. Chakraborty

10. Phytoextraction of Lead-Contaminated Soils:
    Current Experience   225
    John Pichtel

    Index   249

Contributors

P.K. Wong
Department of Biology, The Chinese University of Hong Kong, Shatin, NT, Hong Kong SAR, CHINA.

Patrick K. Jjemba
Biological Sciences Department, University of Cincinnati, P.O. Box 210006 Cincinnati, OH 45221-0006, USA.

Flynn Picardal
Indiana University, Bloomington, IN, 47405, USA.

D. Craig Cooper
Idaho National Laboratory, Idaho Falls, ID, 83415, USA.

Iqbal Ahmad
Department of Agricultural Microbiology, Aligarh Muslim University, Aligarh-202 002, INDIA.

S. Hayat
Department of Botany, Aligarh Muslim University, Aligarh-202 002, INDIA.

A. Ahmad
Department of Botany, Aligarh Muslim University, Aligarh-202 002, INDIA.

A. Inam
Department of Botany, Aligarh Muslim University, Aligarh-202 002, INDIA.

Samiullah
Department of Botany, Aligarh Muslim University, Aligarh-202 002, INDIA.

Galina Koptsik
Soil Science Faculty, Moscow State University, Moscow 119992, RUSSIA.

Steve Lofts
Centre for Ecology and Hydrology, Lancaster Environment Centre,
UNITED KINGDOM.

Elizaveta Karavanova
Soil Science Faculty, Moscow State University, Moscow 119992, RUSSIA.

Natalia Naumova
Institute of Soil Science and Agrochemistry, Russian Academy of Sciences,
Siberian Branch, RUSSIA.

Michiel Rutgers
Laboratory of Ecotoxicology, National Institute for Public Health and the
Environment, THE NETHERLANDS.

Andrew S. Ball
Department of Biological Sciences, University of Essex, Wivenhoe Park,
Colchester CO4 3SQ, UK.

Zaki A. Siddiqui
Department of Botany, Aligarh Muslim University, Aligarh-202 002 INDIA.

Lamabam P. Singh
Department of Botany, Aligarh Muslim University, Aligarh-202 002 INDIA.

K. Chakrabarti
Institute of Agricultural Science, Calcutta University, 35, Ballygunge Circular
Road, Kolkata – 700019, West Bengal, INDIA.

P. Bhattacharyya
Dept. of Geology & Geophysics, Indian Institute of Technology,
Kharagpur-721302, West Bengal, INDIA.

A. Chakraborty
Department of Agronomy, Bidhan Chandra Krishi Viswavidyalaya, P.O.
Mohanpur, Dist: Nadia-741242, West Bengal, INDIA.

John Pichtel
Ball State University, Natural Resources and Environmental Management,
Muncie, IN 47306, USA.

CHAPTER 1

Practical Issues of Land Application of Biosolids

P.K. Wong

Department of Biology, The Chinese University of Hong Kong,
Shatin, NT, Hong Kong SAR, CHINA.

Treatment of domestic and industrial sewage generates substantial amount of effluent and sludge. The improper disposal of sewage sludge or untreated sewage causes serious environmental problems. Application of untreated sewage or sewage sludge as fertilizer or conditioner for soil has been practiced in many countries. However, accumulation of toxic substances, which are present in sewage or sewage sludge, by vegetations poses potential threat to the living organisms in the applied area. Major toxicants in sewage and sewage sludge can be classified as inorganic and organic components. Metal ions are one of the major toxic inorganic components in sewage and sewage sludge and draw most of the concerns due to its persistence and toxicity. In this chapter, the pre-treatment of sewage and sewage sludge by various physical, chemical and biological methods to remove toxic metal ions before application is introduced. The feasibility and potential of these methods in large-scale application are discussed.

APPLICATIONS OF BIOSOLIDS AS A FERTILIZER OR SOIL CONDITIONER

What are Biosolids

The Section 405 (e) of the Clean Water Act (CWA) prohibits any person from disposing sewage sludge generated from any publicly or privately owned sewage treatment works (POTWs) to disposal sites (EPA, 2000b) without proper treatment. Sewage sludge is a nutrient-rich (containing both organic and inorganic compounds) solid material resulting from the treatment of municipal sewage. After appropriate treatment and processing, the residuals can be used as a fertilizer to improve and maintain the productivity of soil and stimulate growth of vegetation including crops or flora (EPA, 1993). These treated solid residuals are termed "biosolids".

How are Biosolids Produced

In the United States, the production of biosolids can begin before the generation of sewage sludge since federal regulation (e.g. the Part 503 Rule which will be discussed in the following sections) requires pre-treatment of the incoming sewage to remove any hazardous materials before treatment by POTWs (EPA, 1994). The typical pre-treatment phase of a conventional sewage treatment works includes physical, chemical and biological processes, which remove non-degradable solids and degrade organic compounds in sewage (EPA, 1994; 1995). There are at least two stages in biological sewage treatment. The first is called primary treatment, in which physical processes such as filtering and settling are used to remove both large and small particle size-solids, respectively, in the incoming raw sewage. Then the sewage is passed on to secondary treatment in which biological processes including: (1) aerobic degradation of organic compounds suspended from primary treatment; and (2) anaerobic degradation of small particle size-solids from primary treatment and excess solids (microbial cell mass and organic compounds that cannot be degraded aerobically). The sludge resulting after anaerobic degradation is the solid by-product of biological sewage treatment processes (EPA, 1993).

Sewage sludge contains high levels of organic compounds and significant amounts of inorganic compounds (most being cationic and anionic species) that can serve as nutrient sources for plants (EPA, 2000d; National Research Council, 2002). Thus, recycling of sewage sludge to land is a sustainable and environment friendly approach to biosolids management (EPA, 1993; 2000d; Andreoli et al., 2002). However, high levels of microbes (some being pathogenic) occur in the sludge, some of the organic components cannot be used directly by plants on the recipient soils, and some organic and inorganic compounds are even toxic to plant species; therefore, treatment(s) to convert sewage sludge to utilizable and non-hazardous biosolids are required by regulations (EPA, 1993; 2000b; 2000d). Treatment options will be discussed in the following sections (Brown et al., 1997a; 1997b; Fang and Wong, 1999; EPA, 2000a).

Practical uses of Biosolids

Due to the high levels of organic and inorganic nutrients such as nitrogen, phosphorus, potassium and trace elements including calcium, copper, iron, magnesium, manganese, sulfur and zinc, biosolids can be used as a fertilizer to improve and maintain soil productivity in agricultural or abandoned sites (Barbarick *et al.*, 1998; Gove *et al.*, 2001, Qiao *et al.*, 2003; Su and Wong, 2003). Land application involves the application of biosolids to land either to condition the soil or to fertilize crops or other vegetation. In the US, there are small- and large-scale applications of biosolids to farmland, gardens, parks and reclaimed mine sites throughout the 50 states (EPA, 1993; 2000d). The application of biosolids not only solves the problem of disposal, its application also reduces the need for chemical fertilizers (EPA, 1993). Nearly 50% of sewage sludge generated in US from POTWs is currently being land applied (EPA, 1993; 2000d). More specifically, sites for biosolids application include: (1) non-public contact sites such as agricultural lands, forests and reclamation sites (e.g. mine sites), and (2) public contact sites such as public parks, plant nurseries, roadsides, golf courses, lawns and home gardens (EPA, 1993; 2000d; 2003; Water Environmental Research Foundation, 2004).

Although the major sites of biosolids application are agricultural lands and gardens, biosolids are also successfully used in the reclamation of abandoned mine sites to establish sustainable vegetation (Illera *et al.*, 2000; Abbott *et al.*, 2001; Andreoli *et al.*, 2002). The organic and inorganic nutrients and matrix provided by biosolids stabilize soil pH, reduce the bioavailability of toxic substances often found in disturbed mine soils, and also regenerates the soil layer to support pioneer vegetation (Andreoli *et al.*, 2002). The use of biosolids to reclaim mine sites usually requires higher rates and longer periods of application than those for agricultural lands. (Abbott *et al.*, 2001).

THE IDENTIFICATION AND CHARACTERIZATION OF PROBLEMATIC SUBSTANCES IN BIOSOLIDS

The Safety of Land Application of Biosolids – EPA Part 503 Rule

The Part 503 Rule consists of seven elements designed to protect human health and the environment from the use of biosolids (EPA, 1994; 1995; 2000b). These elements are:

- General requirements
- Numerical limits for certain pollutants
- Management practices
- Operational standards
- Monitoring
- Record keeping
- Reporting

In order to minimize pollution problems caused by biosolids application, 22 priority pollutants were selected and listed in the Part 503 Rule (Table 1.1). These pollutants can be categorized into groups according to their chemical properties. Their relative levels in biosolids are provided in table 1.2. In the Part 503 Rule, the ceiling concentrations of these pollutants are regulated, as are the application to different types of sites, modes of application, and types of biosolids applied (Table 1.3). In general, regulated pollutants and other components in biosolids can be classified into the following:

Table 1.1 Pollutants selected for potential regulation for land application of biosolids under the EPA Part 503 Rule (EPA, 1993; 1994; 1995; 2000b)

Pollutant	Compounds
Inorganic	
Metals	Arsenic (III)
	Cadmium (II)
	Chromium (III and IV)
	Copper (II)
	Lead (II)
	Mercury (I and II)
	Molybdenum (II)
	Nickel (II)
	Selenium (II)
	Zinc (II)
Organic	
Amines	n-Nitrosodimethylamine
Chlorinated alkenes	Hexachlorobutadiene
	Trichloroethylene
Chlorinated aromatic hydrocarbons	DDT (DDD and DDE)
	Hexachlorobenzene
	Polychlorinated biphenyls
Chlorinated cyclic hydrocarbons	Aldrin
	Dieldrin
Chlordane	Heptachlor
	Lindane
	Toxaphene
Polycyclic aromatic hydrocarbons	Benzo(a)pyrene

Inorganic Nutrients

Nitrogen and phosphorous are essential for plant growth. The levels of nitrogen and phosphorus in biosolids are comparatively higher than those in other organic soil additives (O'Connor et al., 2004). The use of biosolids

Table 1.2 Physical and chemical properties of EPA listed pollutants in biosolids

Pollutant	CAS no.	Carcinogenicity (IARC[a] Cancer Review Group)	Usage	Level in biosolids
Aresenic (II)	7440-38-2	Yes (Group 1)	Hardening metals	Low
Cadmium (II)	7440-43-9	Yes (Group 2A)	Electroplating	Low
Chromium (III, IV)	7440-47-3	Yes (Group 3)	Alloying and electroplating	High
Copper (II)	7440-50-8	No	Electroplating	High
Lead (II)	7439-92-1	Yes (Group 2B)	Batteries	Low
Mercury (I, II)	7439-97-6	No	Catalyst	Low
Molybdenum (II)	7439-98-7	No	Steel manufacturing	High
Nickel (II)	7440-02-0	Yes (Group 1)	Alloying and electroplating	High
Selenium (II)	7782-49-2	Yes (Group 3)	Electronic	Low
Zinc (II)	7740-66-6	No	Alloying and electroplating	High
Aldrin	309-00-2	Yes (Group 3)	Insecticide	Low
Benzo(a)pyrene	50-32-8	Yes (Group 2B)	No specific use	
Chlordane	57-74-9	Yes (Group 3)	Insecticide	Low
DDT (DDD, DDE)	50-29-3	Yes (Group 2B)	Insecticide	Low
Dieldrin	60-57-1	Yes (Group 3)	Insecticide	Low
Heptachlor	76-44-8	Yes (Group 3)	Insecticide	Low
Hexachlorobenzene	118-74-1	Yes (Group 2B)	Wood preservative	Low
Hexachlorobutandiene	87-68-3	Yes (Group 3)	Heat-transfer liquid	Low
Lindane	58-89-9	Yes (Group 2B)	Insecticide	Low
n-Nitrosoamine	62-75-9	Yes (Group 2B)	Cosmetics and antifreeze	Low
Polychlorinated biphenyls	1336-36-3	Yes (Group 2A)	Transformer oil	Low
Toxaphene	8001-35-2	Yes (Group 2B)	Insecticide	Low
Trichloroethylene	79-01-6	Yes (Group3)	Degreasing	Low

[a]IARC = International Agency for Cancer Research

provides adequate quantities of nitrogen (e.g. $NH4^+-N$ and $NO3^--N$) and phosphorus ($PO4^{3-}-P$), which are the most favorable inorganic forms for plant growth. Unfortunately, the C:N:P ratio in biosolids is not in adequate balance, because P levels are higher than those for C and N (Shober and Sims, 2003). Therefore, long-term application of biosolids, without any supplement of C and N, may lead to accumulation of excess P, which can cause problems to plants grown in treated soil.

Table 1.3   Metallic pollutant limits within biosolids for land application (EPA, 1994)

Metal	Ceiling concentration limits for all biosolids applied to land (mg/kg)[a]	Pollutant concentration limits for EQ and PC biosolids (mg/kg)[a]	Cumulative pollutant loading rate limits for CPLR biosolids (kg/he)	Annual pollutant loading rate limits for APLR biosolids (kg/he/365-days)
Arsenic	75	41	41	2.0
Cadmium	85	39	39	1.9
Chromium	3,000	1,200	3,000	150
Copper	4,300	1,500	1,500	75
Lead	840	300	300	15
Mercury	57	17	17	0.85
Molybdenum[b]	75	–	–	–
Nickel	420	420	420	21
Selenium	100	36	100	5.0
Zinc	7,500	2,800	2,800	140
Applies to:	All biosolids that are land applied	Bulk biosolids and bagged biosolids	Bulk biosolids	Bagged biosolids[c]
From Part 503	Table 1, Section 503.13	Table 3, Section 503.13	Table 2, Section 503.13	Table 4, Section 503.13

[a]Dry-weight basis

[b]As a result of the February 25, 1994 Amendment to the Rule, the limits for molybdenum were deleted from the Part 503 Rule pending EPA reconsideration.

[c]Bagged biosolids are sold or given away in a bag or other container

EQ = Exceptional quality (biosolids that meet low-pollutant and Class A pathogen reduction (virtual absence of pathogens) limits and have a reduced level of degradable compounds that attract vectors.

PC = Pollutant concentration (biosolids that meet the same low-pollutant concentration limits as EQ biosolids, but only meet Class B pathogen reduction and/or are subjected to site management practices rather than treatment options to reduce vector attraction properties).

CPLR = Cumulative pollutant load rate (biosolids typically exceed at least one of the pollutant concentration limits for EQ and PC biosolids but meet the ceiling concentration limits).

APLR = Annual pollutant load rate (biosolids that are sold or given away in a bag or other containers for application to the land that exceed the pollutant limits for EQ biosolids but meet the ceiling concentration limits).

Organic Compounds

Most organic compounds are degraded by biological processes during aerobic and anaerobic sewage treatment (Chaney *et al.*, 1996). However, some recalcitrant organic compounds, especially synthetics such as pesticides, dyes and substances used in polymer industries, accumulate in sewage sludge. Although the levels of most of these compounds are low in sewage sludge and biosolids (Table 1.2), reports indicate that levels of polychlorinated biphenyls (PCBs) and detergents are higher than in conventional irrigation water and soil additives (EPA, 1994; 1995). These two classes of synthetic organics are highly toxic to biota and require special attention during biosolids application. In addition, although many synthetic compounds occur at low levels in biosolids, their toxicities may be extremely high and require careful handling and monitoring during application (EPA, 1995). The quality standards, protocol of use and storage, and monitoring during and after land application of biosolids are explicitly listed in the Part 503 Rule (EPA, 1993; 1994; 2000d). The Rule applies to those individuals who prepare, store, trade and apply biosolids.

Pathogens

All biosolids that are used or stored prior to application must be pre-treated to meet the Part 503 Rule for Class A or Class B pathogen density limits (EPA, 1994). Class A biosolids include sewage sludges treated to result in non-detectable pathogens, while Class B biosolids are those treated to result in more than 99% of pathogens inactivated (EPA, 1993; 1994; 2000d). However, the use of Class B biosolids for land application is under an additional set of regulations such as restricted access to application sites to protect humans and animals from infection, protection of water sources from potential contamination by residual pathogens, and proper site selection and management for land application (EPA, 1994). For instance, soils having high clay and organic matter contents may physically filter, adsorb and immobilize pathogens, while sandy soil is porous and pathogens can readily leach to groundwater, and thus is not suitable for application of Class B biosolids (EPA, 1995).

Metals

Certain metals occur in sewage sludges in significantly high concentrations, since metals cannot be degraded by biological treatment process(es) (Shanableh and Ginige, 1999; Bickers et al., 2003). Most metals accumulate in sludge either by adsorption onto microbes or to other solid matter in sewage. Most metals are tightly bound to biosolids components and are not highly water-soluble (EPA, 1995; Solan et al., 1998); thus, metal contamination from land application should not create serious pollution of groundwater or aquatic ecosystems close to application sites, assuming adequate soil conservation practices. However, any change of soil physicochemical properties on biosolids-applied land could result in the release of bound metals into soil and water (Chen et al., 2003). Regular monitoring of metal levels in environments close to biosolids application sites is highly recommended.

Alkaline Stabilization of Sewage Sludge

Both Class A (no detectable pathogens) and Class B (reduced pathogens) biosolids are suitable for land application, but additional requirements are necessary for the use of Class B biosolids for land application (EPA, 2000a). The requirements listed in the Part 503 Rule are: to limit public access to the application site, limiting livestock grazing, and controlling crop harvesting schedules. Alkaline stabilization of sewage sludge is one of the most effective methods to generate biosolids that meet the requirements of pathogen limits for land application. The method stabilizes sewage sludge by adding alkaline materials such as lime to raise pH to a level that is not suitable for the growth of microbes including pathogens (EPA, 2000a; Su and Wong, 2003).

To produce Class A biosolids, sludge pH must be maintained at or above 12 for at least 72 h, with a temperature of 52°C maintained for at least 12 h, or the temperature is maintained at 70°C for at least 30 min during the 72 h incubation period (EPA, 2000a). The high temperature facilitates the inactivation of pathogens. The monitoring of fecal coliforms in treated biosolids is required to ensure elimination of pathogens.

The materials used for alkaline stabilization include hydrated lime (calcium hydroxide), quicklime (calcium oxide) and fly ash (Brown et al., 1997a; 1997b; Fang and Wong, 1999; EPA, 2000a; Wong et al., 2002). Quicklime is the most commonly used material for alkaline stabilization due to its high heat of hydrolysis and efficient pathogen reduction (EPA, 2000a). Fly ash is occasionally used due to its low cost; such utilization may serve as an alternative to ash disposal (Abbott et al., 2001). Alkaline stabilization enhances the ability of biosolids to improve soil properties such as pH, structure and water-holding ability (Brown et al., 1997a; 1997b; Wong et al., 2002).

Land application of biosolids, especially for long-term operation, leads to increase in soil metal concentrations (Brwon et al., 1997b; Pierce et al., 1998). The alkaline pH of biosolids will often increase soil pH and will subsequently maintain metals in insoluble (i.e. non-bioavailable) forms in the applied area (Brown et al., 1997a). However, once the physicochemical conditions of the soil are altered, these metals can be released from soil and will pose a potential threat to local ecosystems.

Other Treatments to Produce Class A and Class B Biosolids

Other than the alkaline stabilization alternative methods also exist to meet the requirements for Class A biosolids (EPA, 2000a):

- Thermal treatment

- Treatment in a high pH-high temperature process
- Treatment by other processes which can reduce enteric viruses and viable helminth ova
- Treatment in one of the "Processes to Further Reduce Pathogens" (PFRP) recommended by the EPA

For Class B biosolids, there are three alternatives to meet the requirements set by the Part 503 Rule (EPA, 1994; 2000a):

- Monitoring of indicator organisms such as fecal coliforms at the time of biosolids use
- Treatment in one of the "Processes to Significantly Reduce Pathogens" (PSRP) recommended by EPA
- Treatment in a process equivalent to one of the PSRP.

However, none of the alternatives for treating Class A or B biosolids reduce metal levels. Thus, there remains the potential threat of metals to sites receiving biosolids over the long-term.

METAL CONTAMINATION FROM LONG-TERM LAND APPLICATION OF BIOSOLIDS

Problems of Metals in Biosolids

Among all the components of biosolids, metals occupy a substantial portion on a dry weight basis (Table 1.1). One of the reasons for the high metal levels is that no specific process is feasible during biological treatment for their removal. Another reason is that metals persist even after chemical or biological treatments, which are directed primarily toward the inactivation of pathogenic organisms and their transmission vectors and removal of toxic organic compounds (EPA, 2000a). The ratio of metals (either in free or bound form) to dry mass of biosolids is comparatively higher than that in raw sewage sludge. Thus the removal of metals from biosolids before application is crucial to reduce the risks posed to local ecosystems.

Reviews Conducted by EPA on the Part 503 Rule Since 1993

Section 405 (d) of the Clean Water Act (CWA) required the EPA to conduct a two-round review of sewage sludge regulations (EPA, 1996). In the first round, EPA was required to establish numerical limits and management practices for toxicants present in sludges based on the existing information on their toxicity, persistence, concentration, mobility, or potential exposure. EPA was required to conduct a second round of reviews on toxicants which were not regulated in the first round, although still likely to impart adverse effects on public health or the environment (EPA, 1996).

Round One Review

In the Round One review, EPA listed 803 chemicals considered hazardous to human health (EPA, 1996). These chemicals were screened with available data on their effects to humans, plants, domestic animals, wildlife and aquatic animals, and frequency of occurrence in biosolids. Two hundred chemicals were identified to be of concern in biosolids (EPA, 1996). A panel of scientists then selected 40 chemicals of potential concern for evaluation by EPA (Water Environmental Research Foundation, 2004). The selected chemicals met two criteria: (1) it has measured concentrations in US sewage sludge based on the published literature or had been measured in the 1989 National Survey of Sewage Sludge (NSSS), and (2) it has a human health benchmark (EPA, 1996; Water Environmental Research Foundation, 2004). A probabilistic hazard assessment model with appropriate, conservative assumptions was subsequently used to analyze the 40 "short-listed" chemicals by EPA. Two questions were addressed in the assessment:

- Which environmental pathways are of concern?
- What is the potential hazard associated with each pollutant?

After this screening process, 22 toxicants were selected for regulation and listed in the Part 503 Rule (Table 1.1). The Rule was published by EPA in 1993 after the Round One review.

Round Two Review

The Round Two review was initiated after the 1993 issuing of the Part 503 Rule (EPA, 1996). EPA considered a preliminary list of 411 hazardous chemicals identified by a risk assessment and pathways in the National Survey of Sewage Sludge conducted in 1990. These chemicals were not included in the Part 503 Rule. A total of 254 chemicals were eliminated from consideration because they were not detected in sewage sludge, while another 69 were eliminated due to their occurrence in less than 10% of all sludge samples analyzed. Pollutants which occurred in more than 10% of samples but with insufficient toxicity data were also eliminated. Table 1.4 lists the candidate chemicals for Round Two review (EPA, 1996). Finally, 31 chemicals including 29 specific congeners of PCDDs, PCDFs and coplanar PCBs (each regarded as a distinct group of chemicals) were adopted by EPA for consideration in the regulation of biosolids use and disposal. Among these 31 chemicals, 10 are metals.

Table 1.4   Candidate pollutants for Round Two regulations[a] (EPA, 1996)

Acetic acid (2,4-dichlorophenoxy)	Methylene chloride
Aluminum[b]	Nitrate
Antimony	Nitrite
Asbestos[c]	Pentachloronitrobenzene
Barium	Phenol
Beryllium	Polychlorinated biphenyls (coplanar)
Bis(2-ethylhexyl)phthalate	2-Propanone
Boron	2-(2,4,5-trichlorophenoxy)-propionic acid
2-Butanone	Silver
Carbon disulfide	Thallium
p-Cresol	Tin
Cyanides (soluble salts and complexes)	Titanium
Dioxins and dibenzofurans	Toluene
Endosulfan-II	2,4,5-Trichlorophenoxyacetic acid
Fluoride	Vanadium
Manganese	

[a]Pollutants detected at a frequency of at least 10% with human health and/or ecological toxicity data available.

[b]Aluminum does not have human health or ecological toxicity data available but is included because of its potential for phytotoxicity.

[c]Asbestos was not tested in the NSSS but is toxic, persistent, and can occur in sewage sludge.

Reviews Conducted by the National Research Council on the Part 503 Rule in 2002

EPA invited the National Research Council (NRC) to evaluate the scientific basis of EPA's current regulations and standards for chemical and microbial pollutants (pathogens) in sewage sludge to be used for land application (National Research Council, 2002). EPA specifically asked NRC to focus on the adequacy and appropriateness of the risk assessment methods and data used to prepare the regulatory requirements to protect human health (EPA, 2003). The major tasks for the NRC were to:

- Review the risk assessment methods and data used to establish concentration limits for chemical pollutants in biosolids to determine whether they are the most appropriate approaches.
- Review the current standards for pathogen reduction or elimination in biosolids and their adequacy for protecting public health.
- Explore whether approaches for conducting pathogen risk assessment can be integrated with those for chemical risk assessment.

In 2002 NRC convened a committee to conduct an independent evaluation of the EPA technical methods and approaches (National Research Council, 2002). The report, entitled "Biosolids Applied to Land: Advancing Standards and Practices", was prepared by the NRC and submitted to EPA in April 2003 (EPA, 2003). The main concerns cited in the report include:

- A lack of exposure and health information of exposed populations.
- Reliance on outdated risk assessment methods.
- Reliance on outdated characterization of sewage sludge.
- Inadequate programs to ensure compliance with biosolids regulation.
- Lack of resources devoted to EPA's biosolids program.

NRC added specific recommendations for chemical and pathogen standards and suggested that EPA consider 14 major exposure pathways (nine of which involve exposure to humans), and proposed other models to replace highly exposed individuals (HEI) in the risk assessment (National Research Council, 2002). In addition, NRC recommended to EPA that more research be conducted to verify that management control was adequate to maintain minimal exposure concentrations over an extended period (National Research Council, 2002).

In response to the NRC's recommendations, EPA, in April 2003, presented responses and a planned strategy by specific categories: (1) survey, (2) exposure, (3) risk assessment, (4) method development, (5) pathogens, (6) human health studies, (7) regulation activities and (8) biosolids management (EPA, 2003). In addition, EPA also invited the Water Environmental Research Foundation (WERF) to hold meetings to solicit views of representatives of stakeholders, government agencies and academics (Water Environmental Research Foundation, 2004). EPA then weighed several factors in determining its final action plan to review the Part 503 Rule: (1) major concerns presented in public comments received in the April 2003 report, (2) the findings of the WERF "Biosolids Research Summit" (introduced in following section) in July 2003, (3) EPA's existing research commitments in response to areas in the NRC report, and (4) feasibility of responding to specific areas given available resources. However, no specific section in the NRC report and EPA response(s) stressed the metals contamination problem in long-term land application of sludges.

Biosolids Research Summit Conducted by the Water Environmental Research Foundation, 2003

One of the major suggestions in the NRC review of the Part 503 Rule was that EPA solicits views of stakeholders who are involved in the preparation or application of biosolids and the public, especially those affected by land application. In 2003, EPA requested the WERF to hold several regional meetings and a final Biosolids Research Summit in July 2003 to follow up the findings of the 2002 NRC study (Water Environmental Research Foundation, 2004). There were four objectives of the Summit:

- To provide a forum for free and open exchange of information on how scientific research can address the concerns and interests of stakeholders involved in land application of Class A or Class B biosolids.
- To develop prioritized "Research Concepts" for recommendation to WERF, EPA or other organizations or agencies involved in monitoring or regulating the application of biosolids.
- To develop a set of principles and recommendations for the above-mentioned organizations and agencies to fund, manage and review scientific research projects.
- To develop a clear statement from the organizations and agencies as a basis for a policy for the participation of scientists and stakeholders to these recommended research concepts.

There were 31 prioritized Research Concepts identified at the Summit (Table 1.5). Those ranked 3, 15, 26, 28 and 29 are related to the analysis and risk of chemicals including metals in biosolids, while Research Concepts 3, 5, 9, 10, 14, 15, 26, 28 and 29 are related to analysis and risk of organic chemicals. The prioritization of these Research Concepts by the government, academic and public sectors indicates that the adverse effects of metals of the sites receiving long-term application of biosolids are underestimated.

EPA Response to the NRC Review

EPA has planned 14 near-term projects for 2004-2005 in order to strengthen the biosolids management program. Specific goals are to improve the ability to: (1) measure pollutants of interest, (2) determine the risk posed by contaminants identified as potentially hazardous, (3) bring various stakeholder groups together via workshops to develop a national incidence tracking system to ultimately determine health effects following land application of biosolids, (4) better understand and characterize the odors, volatile organic chemicals, and bioaerosols emitted from land application sites, (5) better understand the effectiveness of sewage sludge processes and management practices to control pathogens, (6) improve EPA inspection and compliance initiatives, and (7) improve stakeholders' involvement in EPA's biosolids program (EPA, 2003). Project titles are:

1. Biennial Review Under Clean Water Act Section 405 (d) (2) (C).
2. Compliance Assistance and Enforcement Actions.
3. Methods Development, Optimization, and Validation for Microbial Pollutants in Sewage Sludge.

Table 1.5   Priority Research Project Concepts identified at the WERF Research Summit, 2004 (Water Environmental Research Council, 2004)

Rank	Project Concept
1	Rapid incident response study (e.g. outbreak and retrospective case studies, case control study, pilots).
2	Targeted characterization of pathogens in sludge and biosolids (all stages). Variation in treatment: what pathogens are in Class A vs. Class B and ultimate fate in the field.
3	National survey of constituents of concerns in biosolids.
4	Characterization of bioaerosols associated with land applied biosolids.
5	Identify the odor compounds emitted by sludge in the various stages from generation to end use, and specify their sensory potencies and mechanisms of generation and release.
6	Cost-benefit analysis of management options for sludge/biosolids use and disposal.
7	Evaluate the effectiveness of current 503 and other management practices (e.g. BMPs at state and local levels, Manual of Good Practice) of biosolids management and regulatory compliance.
8	Evaluation of the effectiveness of recommended management practices in minimizing pollutant transport from biosolids amended sites.
9	Conduct influent-to-effluent evaluation of treatment processes to reduce or minimize odor generation through process optimization, including investigating additives to control odor.
10	Evaluate emerging and existing treatment technologies (for pathogens, vector attraction reduction, odors, and endocrine disruptor compounds) to achieve desired product quality.
11	Community comparisons (looking at reported health impacted communities and no impact reported communities: all have land application of biosolids/treated sewage sludge).
12	Origins of differences in perception of risks regarding land application of biosolids/treated sludge.
13	Quality of Life Study; Short term study to determine the impact of land application of biosolids on surrounding communities and on those who have health concerns that may be associated with this practice.
14	Odor and human health outcome study.
15	Human and ecological risks from long term land application of chemical contaminants in biosolids and other organic byproducts: changes in bioavailability over time.
16	Evaluation of occurrence of emerging pathogens.
17	Analysis of environmental justice implications of land application of biosolids/ treated sludge.
18	Estimating setback needs to protect sensitive areas such as water wells and surface water from intrusion of pathogens.

Contd.

Table 1.5   Contd.

Rank	Project Concept
19	Development of dose-response relationships for pathogens in biosolids and other organic residuals (including via inhalation) to allow for QMRA from land application.
20	Case studies of local, state, and regional government decisions and involvement in land application.
21	Understanding risks of mixtures, multiple stressors, and discontinuous/time-varying exposure.
22	Sampling strategies for field-testing of indicators and pathogens (including bioaerosols).
23	Development of new methods to detect pathogen classes.
24	Evaluate the possibility of using E.coli in place of fecal coliform determination in the 503 regulation.
25	Analysis of past stakeholder input, and development of better methods for broader involvement by stakeholders.
26	New ecological risk assessment methods to evaluate impact of biosolids-treated land development and use.
27	Develop an absolute standard for vector attraction reduction.
28	Protocol for characterizing fate, transport, and toxicity of existing and emerging chemicals of concern in biosolids.
29	Quantification of source exposure factors (including airborne) of chemical and microbial contaminants from land applied biosolids to allow for risk assessment.
30	White paper to compare land application to other end uses including incineration, landfill, green energy, and other value-added products. Compare: environmental impacts/risk, sustainability indexes, public perception, cost effectiveness, and lifecycle analysis.
31	Risk perception and development of acceptable risk levels for affected populations.

4. Field Studies of Application of Treated Sewage Sludge.
5. Targeted National Survey of Pollutants in Sewage Sludge.
6. Participate in an Incident Tracking Workshop.
7. Conduct Exposure Measurement Workshop.
8. Assess the Quality and Utility of Data, Tools and Methodologies to Conduct Microbial Risk Assessments on Pathogens.
9. Support Pathogens Equivalency Committee.
10. Development and Application of Analytical Methods for Detecting Pharmaceutical and Personal Care Products in Sewage Sludge.
11. Publish the Proceedings of EPA-USDA Workshop on Emerging Infectious Disease Agents and Issues Associated with Animal Manures, Biosolids, and Other Similar By-products.
12. Support "Sustainable Land Application Conference".

13. Review Criteria for Molybdenum in Land-applied Treated Sewage Sludge.
14. Improve Stakeholder Involvement and Risk Communication.

Similar to the WERF report, the 14 projects proposed by EPA to be initiated in 2004-2005 are mainly focused on pathogens associated with biosolids, the appropriateness and development of risk assessment methods, and arousing public involvement and awareness of the safe use of biosolids for land application (EPA, 2003). The danger of metals contamination for sites having long-term biosolids application and the method(s) to reduce metal contents in biosolids are not major issues in these projects. Only Projects 5 and 13, respectively, are indirectly and directly related to metals in biosolids for land application. EPA and other government agencies should reconsider allocating the appropriate resources to address metal contamination problems resulting from long-term land application of biosolids.

SOLUTIONS TO METAL CONTAMINATION FROM LONG-TERM LAND APPLICATION OF BIOSOLIDS

Methods to Reduce Metals in Biosolids

Two major approaches are available to reduce metal concentrations in biosolids: to eliminate metals from sewage sludge before treatment to produce Class A or Class B biosolids; or, to remove metals from biosolids after treatment to reduce pathogens, transmission vectors and degradable organics. The former approach is considered more favorable since the removal of metals after treatment to prepare Class A or Class B biosolids is comparatively costly and difficult. Treatment to prepare Class A or Class B biosolids results in most metals becoming tightly bound to biosolids components, and high-cost chemical solubilization of metals is required. In addition, chemical solubilization may lead to the release of toxic organic compounds which cause even more acute pollution problems to land receiving biosolids.

Methods to remove or reduce metals in raw sludge can be physical, chemical or biological. All these methods have been studied and none is considered perfect (Xiang *et al.*, 2000; Bickers *et al.*, 2003). For instance, physical and chemical methods suffer from high operational cost, high energy requirements and low efficiency in concentrating metals. The most problematic issue regarding chemical methods is the generation of solid wastes, which contain extremely high concentrations of metals. Additionally, chemical methods require a costly post-treatment. The use of biological treatment to remove metals from raw sludge also has its share of problems. For instance, biosorption, the use of biomass to adsorb metals, can be completed within a short period of time (usually minutes); however, the metal-laden biomass requires additional treatment to recover the metals in a suitable form (Xiang *et al.*, 2000; Bickers *et al.*, 2003). Bioleaching can overcome most of the above-mentioned problems during physical methods, chemical methods and biosorption, however, the long treatment time restricts the use of bioleaching to those regions which are not space-limited (Xiang *et al.*, 2000; Chen *et al.*, 2003). Furthermore, the leaching solution, containing variable concentrations and species of metals, requires suitable recovery technique(s) to make the operation viable (Xiang *et al.*, 2000; Chen *et al.*, 2003).

Other Alternatives to Manage Metals in Biosolids

The increase in alkalinity of raw sludge by addition of chemicals such as lime creates similar effects as for alkaline stabilization of sludges to eliminate pathogens (EPA, 1994; Brown *et al.*, 1997a; 1997b; Fang and Wong, 1999). Once the pH of raw sludge reaches a certain level, insoluble metal hydroxide(s) form and can be removed from the inter-particle spaces of the sludge. However, effective precipitation only occurs in the presence of sufficiently high concentrations of metals, and the metal sludge generated cannot be completely separated from treated sludge without further treatment to produce Class A or Class B biosolids.

Alkaline wastes such as fly ash, a solid waste with a comparatively high pH resulting from fossil fuel combustion, can serve as an alternate material for sludge metal treatment (Su and Wong, 2003). The use of alkaline fly ash to precipitate metals in raw sludge helps to solve the problem of fly ash disposal and, at the same time, the treatment also reduces metal concentrations in raw sludge. Other sludge management alternatives, such as long-term surface storage (EPA, 2000c) and incineration (EPA, 1993; 1994), also address the metal contamination problem. The potential of using these options will be further discussed in the last section.

Increasingly Stringent Control of Metal Content in Biosolids – the European Communities' Initiative

Biosolids are widely land-applied in European Communities (EC). Due to the large number of countries in the Communities, there are as yet no unified guidelines on the use of biosolids in land application. However, some suggestions regarding appropriate levels of organic and inorganic compounds in biosolids have been published. Ceiling concentrations of metals in biosolids published by EC are listed in Table 1.6. There are two set of ceiling values for seven metals, one set issued in 1986 with levels similar to those

Table 1.6   European Union limit values for concentration of heavy metals in biosolids for use on land (Council of the European Communities, 1986).

Elements	Limit values (mg kg^{-1} of dry matter)	
	Directive 86/278/EEC	Proposed
Cadmium	20–40	10
Chromium	–	1,000
Copper	1,000–1,750	1,000
Mercury	16–25	10
Nickel	300–400	300
Lead	750–1,200	750
Zinc	2,500–4,000	2,500

regulated by the Part 503 Rule (Table 1.3). However, the EC proposed more stringent limits for these metals, which should be implemented no later than 2005. Compared with US limits, much lower concentrations of these metals will be acceptable by the EC countries to use biosolids in land application after 2005. The same or similar approach can be adopted by the US to reduce metal pollution caused by long-term land application.

CONCLUSIONS

How serious is Metal Contamination Caused by Land Application of Biosolids

At present, there are few systematic studies on the impacts of metal contamination at sites undergoing long-term application of biosolids. Most studies have reported results from a few months to years monitoring of metal contents in aquatic and terrestrial environments near sites undergoing land application (Brown et al., 1997a; 1997b; Luxmoore et al., 1999; Illera et al., 2000; Shober and Sims, 2003). EPA is conducting research and monitoring programs on metal levels at various sites to study the effects on organisms and the local environment receiving biosolids over the long term. One such site, which is being monitored continuously, has been receiving biosolids for 17 years (Solan et al., 1998). The results of metals monitoring data will provide more evidence about the safe use of biosolids in land application.

Other Alternatives for Disposal or Treatment of Biosolids

In the Part 503 Rule, requirements apply not only to land application of biosolids, but also to surface disposal and incineration of biosolids (EPA,

1993; 1994). Such methods have not received the same degree of attention as land application.

Surface Disposal

Surface disposal is the placement of biosolids on a parcel of land at high rates for final disposal, rather than using the biosolids to condition the soil or use its nutrients to fertilize a crop (EPA, 2000c). Placing biosolids in a landfill, in a surface impoundment, on a waste pile, or on a dedicated site is considered surface disposal. However, treatment and storage of biosolids are not considered surface disposal.

Storage

Storage of biosolids is the placement of biosolids on land for two years or less, while placement on land for more than two years is considered surface disposal (EPA, 2000c).

Treatment

Treatment is the preparation of biosolids for final use or disposal through activities such as thickening, stabilization or dewatering (EPA, 2000b).

Incineration

Biosolids are combusted with auxiliary fuels including gas, oil, coal and others as a form of treatment (EPA, 2000b). The co-combustion of biosolids in an incinerator with other wastes is not uncommon in the US. Additional regulations are imposed on the incineration of biosolids, such as continuous monitoring of carbon monoxide emissions from incinerators.

Among the above activities, surface disposal and incineration should be given further consideration by regulatory agencies, as metal contamination from long-term land application of biosolids continues to be a serious and unresolved problem. Surface disposal can be modified in order to prevent access of the public and wildlife. Unfortunately, the costs for preparation, operation and maintenance of surface disposal sites are much higher than those of land application sites. However, with improvement in design and operation of surface disposal practices, this alternative should also be reconsidered to meet the increasingly stringent safety measures for land application. The most serious problems of using incineration to treat biosolids are: (1) the potential air pollutants produced; and (2) the source(s) and type(s) of co-fired waste(s). For the first problem, an electrostatic precipitator installed upstream of the incinerator stack (flue) will significantly reduce particulate emissions from the plant. With the addition of appropriate fuel, there is no evidence of the production of toxic air pollutants from combustion of biosolids and co-fired wastes at high temperatures. The source(s) and type(s) of co-fired wastes, is, in fact, the major hurdle regarding incineration of biosolids. However, if land application of biosolids becomes a major cause of pollution and surface disposal is not deemed practical, the use of more costly fossil fuels for incineration of biosolids must be considered.

Hidden Threat(s) of Biosolids

Most of this article has focused on safety issues, particularly metal pollution problems from long-term land application of biosolids. However, many biosolids management activities are not specifically regulated by the Part 503 Rule. These involve the use, storage and treatment of biosolids (Table 1.7). Without proper management and authorized agency(ies) to monitor the safe operation of these activities, the risk of disposal and improper management of biosolids remains. Special attention should be provided by government agencies to protect human health and to prevent pollution of local ecosystems from these management activities.

Table 1.7 Activities excluded from the Part 503 Rule (EPA, 2000b; 2000c)

Part 503 does not include requirement for the following activities:	Applicable Federal Regulation
Treatment of Biosolids Processes used to treat sewage sludge prior to final use or disposal (e.g. thickening, dewatering, storage, heat drying).	None (except for operational parameters used to meet the Part 503 pathogen and vector attraction reduction requirements).
Selection of Use or Disposal Practice The selection of a biosolids use or disposal practice.	None (the determination of the biosolids use or the disposal practice is a local decision).
Incineration of Biosolids with Other Wastes Biosolids co-fired in an incinerator with other wastes (other than as an auxiliary fuel).	40 CFR Parts 60, 61
Storage of Biosolids Placement of biosolids on land for two years or less (or longer when demonstrated not to be a surface disposal site but rather, based on practices, constitutes treatment or temporary storage).	None
Industrial Sludge Sludge generated at an industrial facility during the treatment of industrial wastewater with or without combined domestic sewage.	40 CFR Part 257 if land applied 40 CFR Part 258 if placed in a municipal solid waste landfill.

Contd.

Table 1.7   Contd.

Part 503 does not include requirement for the following activities:	Applicable Federal Regulation
Hazardous Sewage Sludge Sewage sludge determined to be hazardous in accordance with 40 CFR Part 261, *Identification and Listing of Hazardous Waste.*	40 CFR Parts 261-268
Sewage Sludge Containing PCBs [3] 50 mg/kg Sewage sludge with a concentration of polychlorinated biphenyls (PCBs) equal to or greater than 50 milligrams per kilogram of total solids (dry-weight basis).	40 CFR Part 761
Incinerator Ash Ash generated during the firing of biosolids in a biosolid incinerator.	40 CFR Part 257 if land applied 40 CFR Part 258 if placed in a municipal solid waste landfill.
Grit and Screenings Grit (e.g. sand, gravel, cinders) or screenings (e.g. relatively large materials such as rags) generated during preliminary treatment of domestic sewage in a treatment works.	40 CFR Part 257 if land applied 40 CFR Part 258 if placed in a municipal solid waste landfill.
Drinking Water Sludge Sludge generated during the treatment of either surface water or ground water used for drinking water.	40 CFR Part 257 if land applied 40 CFR Part 258 if placed in a municipal solid waste landfill.
Certain Non-domestic Septage Septage that contains industrial or commercial septage, including grease-trap pumpings.	40 CFR Part 257 if land applied 40 CFR Part 258 if placed in a municipal solid waste landfill.

REFERENCES

Abbott, D.E., Essington, M.E., Mullen, M.D., Ammons, J.T. (2001). Fly ash and lime-stabilized biosolids mixtures in mine spoil reclamation: Simulated weathering. Journal of Environmental Quality 30: 608-616.

Andreoli, L.A.R., de Castro, A.G., Picinatto, A.G., Pegorinin, E.S. (2002). Effect of the employments of biosolids as a recovery agent for degraded areas. Water Science and Technology 46: 209-216.

Barbarick, K.A., Ippolitp, J.A., Westfall, D.G. (1998). Extractable trace elements in the soil profile after years of biosolids application. Journal of Environmental Quality 27: 801-805.

Bickers, P.O., Chong, R.C., Bhamidimarri, R., Killick, M.G. (2003). Bioleaching of metals from sludge and acid production under increased metal concentrations. Water Science and Technology 48: 169-176.

Brown, S., Angle, J.S., Chaney, R.L. (1997a). Correction of lime-biosolid induced manganese deficiency on a long-term field experiment. Journal of Environmental Quality 26: 1375-1384.

Brown, S., Chaney, R.L., Angle, J.S. (1997b). Substrate liming and metal movement in soil amended with lime-stabilized biosolid. Journal of Environmental Quality 26: 724-732.

Chaney, R.L., Ryan, J.A., O'Connor, G.A. (1996). Organic contaminants in municipal biosolids: Risk assessment, quantitative pathway analysis, and current research priorities. The Science of the Total Environment 185: 187-216.

Chen, S.Y., Lin, J.G., Lee, C.Y. (2003). Effects of ferric ion on bioleaching of heavy metals from contaminated sediment. Water Science and Technology 48: 151-158.

Council of the European Communities. (1986). Council Directive 86/278/EEC of 12 June 1986 on the Protection of the Environment, and in Particular of the Soil, When Sewage Sludge is Used in Agriculture. Community Legislation in Force. Document 386L0278.

Fang, M., Wong, J.W.C. (1999). Effects of lime amendment on availability of heavy metals and maturation in sewage sludge composting. Environmental Pollution 106: 83-89.

EPA. (1993). Standards for the Use or Disposal of Sewage Sludge (40 Code of federal Regulation Part 503). Office of Wastewater Management, US Environmental Protection Agency, Washington, D.C.

EPA. (1994). A Plain English Guide to EPA Part 503 Biosolids Rule. EPA/832/R-93-003, Office of Wastewater Management, US Environmental Protection Agency, Washington, D.C.

EPA. (1995). A Guide to the Biosolids Risk Assessments for the EPA Part 503 Rule. EPA/832/B-93-005, Office of Wastewater Management, US Environmental Protection Agency, Washington, D.C.

EPA. (1996). Technical Support Document for the Round Two Sewage Sludge Pollutants. EPA/822/R-96-003, Office of Water, Office of Science and Technology, Health and Ecological Criteria Division, US Environmental Protection Agency, Washington, D.C.

EPA. (2000a). Biosolids Technology Fact Sheet. Alkaline Stabilization of Biosolids. EPA/832/F-00-052, Office of Water, US Environmental Protection Agency, Washington, D.C.

EPA. (2000b). Federal Register: February 19 1993. 40 CFR Parts 257, 403 and 503. The Standards for the Use or Disposal of Sewage Sludge. Final Rules. EPA/822/Z-93-001, US Environmental Protection Agency, Washington, D.C.

EPA. (2000c). Guide to Field Storage of Biosolids. EPA/832/B-00-007, Office of Wastewater Management, US Environmental Protection Agency, Washington, D.C.

EPA. (2000d). Land Application of Biosolids. EPA/832/F-00-064, Office of Water, US Environmental Protection Agency, Washington, D.C.

EPA. (2003). Standards for the Use or Disposal of Sewage Sludge; Final Agency Response to the National Research Council Reports on Biosolids Applied to Land and the Results of EPA Review of Existing Sewage Sludge Regulation. EPA/FRL-7605-8, Office of Wastewater Management, US Environmental Protection Agency, Washington D.C.

Gove, L., Cooke, C.M., Nicholson, F.A., Beck, A.J. (2001). Movement of water and heavy metals (Zn, Cu, Pb and Ni) through sand and sandy loam amended with biosolids under steady-state hydrological conditions. Bioresources Technology 78: 171-179.

Illera, V., Walter, I., Souza, P., Cala, V. (2000). Short-term effects of biosolids and municipal solid waste application on heavy metals distribution in a degraded soil under a semi-arid environment. The Science of the total Environment 255: 29-34.

Luxmoore, R.J., Tharp, M.L., Efroymson, R.A. (1999). Comparison of simulated forest responses to biosolids application. Journal of Environmental Quality 28: 1996-2007.

National Research Council. (2002). Biosolids Applied to Land. Advancing Standards and Practices. The National Academic Press, Washington, D.C.

O'Connor, G.A., Sarkar, D., Brinton, S.R., Elliott, H.A., Martin, F.G. (2004). Phytoavailability of biosolids phosphorus. Journal of Environmental Quality 33: 703-712.

Pierce, B.L., Redente, E.F., Barbarick, K.A., Brobst, R.B., Hegeman, P. (1998). Plant biomass and element changes in shrubland forages following biosolids application. Journal of Environmental Quality 27: 789-794.

Qiao, X.L., Luo, Y.M., Christie, P., Wong, M.H. (2003). Chemical speciation and extractability of Zn, Cu and Cd in two contrasting biosolids-amended clay soils. Chemosphere 50: 823-829.

Shanableh, A., Ginige, P. (1999). Impact of metals bioleaching on the nutrient value of biological nutrient removal biosolids. Water Science and Technology 39: 175-181.

Shober, A.L., Sims, J.T. (2003). Phosphorus restrictions for land application of biosolids: Current status and future trends. Journal of Environmental Quality 32: 1955-1964.

Solan, J.J., Dowdy, R.H., Dolan, M.S. (1998). Recovery of biosolids-applied heavy metals sixteen years after application. Journal of Environmental Quality 27: 1312-1317.

Su, D.C., Wong, J.W.C. (2003). Chemical speciation and phytoavailability of Zn, Cu, Ni and Cd in soil amended with fly ash-stabilized sewage sludge. Environmental International 29: 895-900.

Water Environmental Research Foundation. (2004). Proceedings from the Biosolids Research Summit. WERF/03-HHE-1, Consensus Building Institute, Cambridge.

Wong, J.W.C., Xiang, L., Chan, L.C. (2002). pH requirement for the bioleaching of heavy metals from anaerobically digested wastewater sludge. Water Air and Soil Pollution 138: 25-35.

Xiang, L., Chan, L.C., Wong, J.W.C. (2000). Removal of heavy metals from anaerobically digested sewage sludge by isolated indigenous iron-oxidizing bacteria. Chemosphere 41: 283-287.

CHAPTER 2

Bioavailability of Metals and Metalloids in Terrestrial Environments

Patrick K. Jjemba

*Biological Sciences Department; University of Cincinnati,
P. O. Box 210006 Cincinnati, OH 45221-0006, USA.*

The presence of heavy metals and metalloids in soil, sediments and water at concentrations that exceed regulatory limits is of major concern from both public health and environmental perspectives. Sewage sludge is the primary source of heavy metals that are introduced to agricultural land. Furthermore, some metals have the propensity to leach from terrestrial environments into groundwater or migrate to surface water, subsequently impacting drinking water supplies. The concentrations that are acceptable by different regulatory authorities vary substantially. Unlike organic compounds, metals are not degraded but rather only transformed from one form to another. Transformation in relation to factors such as sorption, bond strength, and ionic radius affect the bioavailability of metals. However, these variables are not reflected in the regulatory concentrations that have been established. The toxicity characteristic leaching procedure has been used to assess the leachability of metals and to determine the best demonstrated achievable remediation technology from an engineering perspective. However, this approach seems inadequate for predicting the bioavailability of metals to a variety of organisms as well as their subsequent concentration in the food chain. The environmental chemistry of individual metals and how their availability is influenced by both biotic and abiotic systems is examined. Options to reduce bioavailability of metals during sewage treatment and application to land, as well as feasible alternatives to remediate contaminated sites are outlined.

SOURCES OF HEAVY METAL AND METALLOID POLLUTION IN THE ENVIRONMENT

Heavy metal and metalloid pollution is a concern in both agricultural and urban areas. It is apparent from Table 2.1 that the ranges deemed acceptable by regulatory authorities in different countries greatly vary. Most of the contamination from metals to agricultural land is associated with sewage sludge (i.e., biosolids) although some metal smelting activities, as well as application of fertilizers, herbicide and pesticides are additional anthropogenic sources (Stehouwer *et al.*, 2000; Dai *et al.*, 2004; Jjemba, 2004). Metal content in biosolids is often dependent on the extent of industrialization of the area from which the sewage is derived, demonstrating that heavy metals originate primarily from industrial sources (Stehouwer *et al.*, 2000).

Current methods used in the treatment of sewage sludge focus on concentrating heavy metals from the aqueous phase (wastewater) onto the sludge biomass fraction (Atkinson *et al.*, 1998; Manios *et al.*, 2003). The high propensity of the solid fraction in wastewater and sludge to sorb metals is attributed to the existing bacterial extracellular polymers and organic material in the sludge. Their concentrations in the final product (biosolids) are controlled to a certain extent by processes in the treatment system. The processes of removing heavy metals from sewage sludge are fairly diverse; for example, in analyses by Stehouwer *et al.* (2000), metals were generally lower in biosolids that were treated using alkaline digestion with either aerobic or anaerobic digestion.

The final product after sewage treatment is typically disposed on to agricultural land, hauled to a landfill, or discharged into waterways. Disposal of biosolids to agricultural land is the prime option in most industrialized and semi-industrialized countries. For example, in the UK, 47% of the 1.12 million tons (dry solids) sewage sludge is applied to land annually (Green *et al.*, 2003). These applications serve to replenish soil nutrients and increase humus content which in turn significantly improves soil structure and water-holding capacity. However, such applications also introduce heavy metals and metalloids for possible uptake by plants and other soil dwellers, subsequently becoming biomagnified through the food chain (Green *et al.*, 2003). Furthermore, some metals have the propensity to leach into groundwater or migrate to surface water, ultimately impacting drinking water (Harvey *et al.*, 2002).

METAL TOXICITY AND BIOAVAILABILITY

When present in excessive quantities, heavy metals and metalloids can reduce the growth of plants (Logan *et al.*, 1997; Lai and Chen, 2004). They also negatively affect the activity of microorganisms, adversely affecting important biological processes such as nitrogen fixation (Smith, 1997), as well as ATP production (Chander and Brookes, 1991), soil enzyme activity (Kuperman and Carrierro, 1997; Naseby and Lynch, 1997), and microbial biomass production (Insam *et al.*, 1996; Pascual *et al.*, 1999; Rost *et al.*, 2001).

In studies by Khan and Scullion (2002), Cu, Zn, and Pb significantly increased the C/N ratio temporarily. The increase was primarily associated with higher concentrations of these metals, which decreased biomass C. The mechanisms of heavy metal toxicity have been widely studied. Mechanisms include binding more strongly to functional sites that are normally occupied by essential metals, thus blocking the essential functional groups of biologically important molecules such as enzymes, changing the conformation of biological molecules (i.e., proteins and nucleic acids), thus disrupting the integrity of entire cells and/or their membrane, making them inactive, decomposing essential metabolites, and changing the osmotic balance around the cells (Jjemba, 2004).

Heavy metals in the biosphere will out-compete other cations in the cytoplasm for sites on intracellular molecules such as proteins, polysaccharides, and nucleic acids. However, the effects of heavy metals on organisms in specific environments are sometimes contradictory (Giller *et al.*, 1998). Thus, even though biosolids are typically associated with the presence of heavy metals, numerous reports of enhanced vegetation growth following their application have been documented. For example, in studies by Pascaul *et al.* (1999), eight years after application of the different rates of biosolids, the percent vegetative cover and plant species diversity were highest in plots that received the highest dose of biosolids, and was lowest in the control plots. The authors attributed these increases to increased total organic carbon content coupled with increased enzyme activity (i.e., urease, ∃-glucosidase, phosphatase) in the biosolids-treated plots. In contrast, the yields of cereals and legumes were consistently reduced over four years in fields where heavy-metal contaminated sludge had been applied more than two decades after application (Bhogal *et al.*, 2003). Similarly, the yield of greengram declined in pots when heavy metals (Cd, Pb, Cu, Zn, Cr and Ni) were added individually or combined in various mixtures (Athar and Ahmad, 2002).

Some of the contradictions regarding the impact of metals on biota can be attributed to the widespread lack of distinction between total metal concentrations and their bioavailability in specific environments. In soils and other media, metals undergo complexation and, like most organic pollutants, are subject to aging processes that subsequently reduce their bioavailability. However, unlike organic pollutants, metals are not decomposed but rather only transformed. Transformation also affects their subsequent bioavailability and mobility in soils. Thus, both metal aging (sorption) and transformation processes have ecotoxicological implications that are not always addressed when metal-polluted sites are being investigated. It is not entirely clear how the concentrations of the metals that are deemed acceptable by various countries (Table 2.1) were ultimately established. However, it is unlikely that metal bioavailability was

Table 2.1 Permissible limits for various heavy metals in agricultural soils in different countries[a]

Jurisdiction	Permissible concentration (mg kg^{-1} soil)								
	Cr	Ni	Cu	Zn	Cd	Pb	Hg	As	B
General benchmark	100	15	10	40	0.1	20	0.01	–	–
Australia[b]	100	60	100	200	1	150	1	20	–
Belgium (sandy soil)	100	30	50	150	1	50	1		
Belgium (clay-loam soils)	150	75	140	300	3	300	1.5		
China (pH<6.5)	600	100	250	500	5	300	5	75	150
China (pH>6.5)	1000	20	500	1000	20	1000	15	75	150
Netherlands	75	30	75	300	1.25	100	0.75	15	
Republic of S. Africa	80	15	100	185	2	56	0.5	2	10
United Kingdom	400[c] (pH6–7)	75	135[c] (pH6–7)	300[d] (pH5–5.5)	3[d] (pH5–5.5)	300			
USA[e]	1200[f]	420	500	2800	39	300			
Sweden	30	30	40	75	0.4	40	0.3		
New Zealand		35	140	300					

[a]Unless indicated all were compiled from Matthews (1996).
[c]Source: Smith (1997)
[e]Stehouwer et al. (1993)

[b]Guidelines are non-mandatory
[d]Source: Green et al. (2003)
[f]Currently not limited

considered when establishing these values. By definition, bioavailability is the degree to which a contaminant in a particular matrix is available for uptake by an organism (Newman and Jagoe, 1994). In the terrestrial environment, the bioavailable metal is the fraction in the soil solution that is readily available for uptake. It is typically a low concentration; for example, the concentrations of Cu, Ni, and Zn in sandy soil solution were a meager 0.04-0.16%, 0.03-0.19%, and 0.02-0.19%, respectively, of the total soil concentrations in a pasture field in New Zealand (Horswell *et al.*, 2003).

Metal bioavailability in terrestrial environments is influenced by several factors such as pH, soil type, redox potential, moisture status, metal species, and level of organic matter. The pH is particularly important as it controls the behavior of metals and several other soil processes. Under alkaline pH conditions, precipitation occurs for most metal precipitates, which generally decreases their bioavailability and subsequent toxicity to microorganisms in the environment (Babich and Stotzky, 1983). Bioavailability of certain metals and their subsequent toxicity to cells is reduced under acidic conditions. In contrast, Oliver (1997) reports that heavy metal cations are more mobile in soil under acidic conditions, subsequently migrating into groundwater. These contradictions clearly demonstrate that generalizations about metal bioavailability are difficult to make, and that specific metals and/or their oxidation states can behave in a widely differing manner. Furthermore, the availability of some heavy metals can also be enhanced by the presence of other metals (Athar and Ahmad, 2002). For example, in studies by Renella *et al.* (2003), the availability of Zn in soil was increased by Cd whereas the availability of Cu was unaffected. Competition between Cd and Zn has also been reported by others (Elzinga *et al.*, 1999). Bioavailability of a particular metal is also affected by that metal's ability to bind on to organic matter, which in turn can protect the metal against biotransformation (Smith, 1997; Baladane *et al.*, 1999). The organic matter-bound metal complex formed can be very stable and unaffected by changes in pH.

In most instances, the extractable amounts of a metal reflect the total concentration of the metal in the matrix rather than the proportion that is bioavailable. A more accurate assessment of the effects of metals on biota demands a consideration of their abundance in the soil solution, which reflects the bioavailable fraction. It is important to realize that the methods of extracting metals from soil vary and actually recover different concentrations of the same metal (Haq *et al.*, 1980). Typically, the total concentration of metals in soils and sediments is determined using hot concentrated (nitric) acid whereas the bioavailable fraction is extracted using 0.1 M $CaCl_2$ (McGrath and Cegarra, 1992); ammonium acetate together with EDTA (Tessier *et al.*, 1979); or acetic acid (Quevauviller *et al.*, 1997). The toxicity characteristic leaching procedure (TCLP) has been widely used by regulatory authorities to determine the fate and risk associated with metals in contaminated environments (Federal Register, 1990). During TCLP, the leaching is conducted with an acidic solution. This assay is designed to measure the leachability of wastes and the risk such wastes pose to groundwater. The TCLP is useful in determining the best demonstrated achievable technology for engineering purposes; however, various research groups have found it inappropriate for predicting the bioavailability of metals in a variety of plants and other organisms as well as their subsequent concentration in the food chain (Berti and Jacobs, 1996; Poon and Lio, 1997; Snyman, 2001; Mercier *et al.*, 2002).

In the terrestrial environment, bioavailability is ideally determined using bioassays by measuring plant uptake (Smith, 1997) or uptake by other soil dwellers such as earthworms. However, bioassays may not always be possible to address concerns over a short period of time. Thus, some metal extraction methods have been demonstrated to correlate well with bioavailability in model organisms. For example, the NH_4NO_3-extractable fraction and the NH_4-EDTA-extractable fraction correlate with the bioavailability of metals in soils (Snyman, 2001; Renella *et al.*, 2003). In other instances, extractions in 5 mM diethylenetriamine pentaacetic acid (DTPA) in 10 mM $CaCl_2$ for 1h with shaking provided metal concentrations that correlated with the concentration of Zn (r=0.62 and 0.76) and Pb (r=0.73 and 0.57) in two diverse earthworm species. However, these extraction procedures were only weakly correlated with Cd (r=0.35 and 0.48) in both earthworm species and with Cu (r=0.42) for one of the species (Dai *et al.*, 2004). These results underscore the fact that different extracts and extraction procedures may not be suitable for all metals of interest even for the same soil. In general, extraction procedures that include a chelating agent effectively predict bioavailability.

A biota-to-soil accumulation factor (BSAF) using invertebrates has been proposed by Cortet *et al.*, (1999), from the equation below:

BSAF= (metal content in the organism/metal content in soil)

The larger the BSAF ratio, the higher the bioavailability of the metal to the organism in question (Fig. 2.1). In the figure, the bioconcentration of metals is lower in *Lumbricus rubellus* than *L. caliginosa* because the former is a litter dweller whereas the latter is a soil dweller.

Also widely used for determining safe levels of metals to organisms are the ecological dose (ED) and ecological concentration (EC) values. For example, the ED_{50} or EC_{50} signify a 50% reduction in the functioning of the basic ecological process under consideration. Both sludge-amended and unamended treatments that received increasing doses of Cd ranging between 25 and 8,000 mg Cd kg^{-1} soil generally had increasing ED values following incubation for 3 hours to 40 days (Moreno *et al.*, 2002). Because ED_{50} and EC_{50} represent such high losses, i.e., a 50% reduction in function, some workers prefer using ED_{10} or ED_5 values instead.

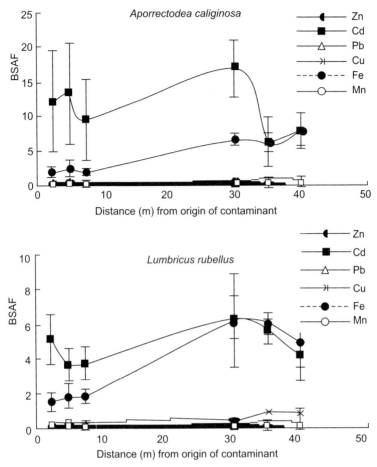

Fig. 2.1   The biota-to-soil accumulation factor (BSAF) of six heavy metals in two earthworm species at a contaminated site (Based on data from Dai *et al.*, 2004 with permission from Elsevier).

RELATIONSHIP BETWEEN CHEMICAL CHARACTERISTICS AND BIOAVAILABILITY

Bioavailability requires an appreciation of the chemical characteristics of individual metals, including how they behave under the influence of other metals and biotic systems (e.g., rhizosphere, microbial cells, invertebrates, etc.). The availability of different inorganic cations differ slightly, some differences following a predictable pattern depending on their grouping in the Periodic table. Some of the chemical attributes that impact the bioavailability of heavy metals in the biosphere are summarized in Table 2.2. Most of these metals are either soft or intermediate acids with low

Table 2.2 Chemical attributes of importance to the bioavailability of various heavy metals in soils[a]

Element	Group[b]	Type[c]	Acidity	Biogenic bond[d]	Ionic radius (pm)	Oxidation states	Other remarks
Cr	VI or 6	TM	Intermediate	No	III is 61.5; VI is 44	II to VI	122 mg/kg in earth's crust. Cr(III) and Cr(VI) are only ones relevant in the environment. Cr(III) is benign and mobile while Cr(VI) is relatively toxic. Essential in low doses.
Mn	VIIB or 7	TM	Hard	No	II is 67; IV is 53; VI is 25; VII is 46	II to VII	1060 mg/kg in the earth's crust. Third most abundant transition metal after Fe and Ti. Essential in low doses.
Ni	VIIIB or 10	TM	Intermediate	Yes	II is 69; III is 56; IV is 48	II to IV	99 mg/kg in earth's crust. Essential in low doses.
Cu	IB or 11	TM	Soft	No	I is 77; II is 73; III is 54	I to III	68 mg/kg in earth's crust. Essential in low doses.
Zn	IIB or 12	TM	Intermediate	No	74	+2	76 mg/kg of earth's crust. Essential in low doses.
Cd	IIB or 12	TM	Soft	No	95	+2	Only 0.16 mg/kg of earth's crust. Low mobility in some soils but high bioavailability.
Hg	IIB or 12	TM	Soft	Yes	119	+1	Only 0.08 mg/kg of earth's crust.
Al	IIIA or 13	M	Soft	No	53	+3	Most abundant metal in earth's crust (8.3% by weight) after O (45.5%) and Si (25.7%)
Pb	IVA or 14	M	Soft	Yes	II is 119; IV is 78	I to IV	13 mg/kg of earth's crust.
As	VA or 15	Md	Soft	Yes	III is 58; V is 46	I to V	Only 1.8 mg kg^{-1} of earth's crust. Mobility in soil was previously underrated. As(III) is toxic; As(V) is benign and less mobile.

[a] Ionic radius and remarks mostly compiled from Greenwood and Earnshaw (1997). The bolded elements are discussed in the text.

[b] Refers to grouping in the periodic table.

[c] M = metal, Md = metalloid, TM = transition metal.

[d] Ability to form biogenic bonds.

electronegativity and a high propensity to polarize. This trait enables the soft acids to compete effectively with hard acids, displacing them and forming more stable cellular complexes. Among the metals listed, only Mn is a hard acid. Hard acids are highly electronegative and do not readily polarize. A majority of the metals listed in Table 2.2 are also transition metals. Transition metals tend to bind ligands strongly as they have a partially-filled d orbital shell.

Some of the heavy metals listed are essential for organisms. They are involved in biological processes such as stabilizing proteins or acting as catalysts, but are required only in trace amounts.

Bioavailability is also affected by sorption bond strength and ionic radius. Most of the heavy metals listed in Table 2.2 have a large ionic radius and several oxidation states. The larger the ionic radius, the less likely a particular metal is able to diffuse into the structural lattice of the soil particles. On the other hand, if the ionic radius is small, the metal will migrate into the micropores of the soil and remain inaccessible to biological targets. It follows from this analogy, therefore, that within the same soil or sediment, bioavailability differs between the respective heavy metals. Nickel, Hg, Pb and As are able to form biogenic bonds with the most abundant element in the environment, i.e., carbon, forming methylated complexes. The chemical attributes highlighted above have ramifications to the bioavailability of heavy metals in the environment as briefly illustrated with four of the most important heavy metals (Cr, As, Cd, Pb) discussed below.

Chromium

Chromium occurs in oxidation states – II to VI but only states 0, III and VI are the most stable. In the terrestrial environment, it exists as a trivalent cation (Cr^{3+}), a trivalent anion (CrO^{2-}), or in two hexavalent forms, i.e., $Cr_2O_7^{2-}$ and CrO_4^{2-}. Only Cr(III) and Cr(VI) are important in the environment, with the former being more benign to biota than the latter. Chromium is widely used in the stainless steel industry because of its significant anti-corrosion properties. It is also used in photocopier toners, leather tanning, textile pigments and in wood preservatives. It is found in soils at concentrations ranging between 10-150 mg kg^{-1} (Adriano, 2001). Small quantities of Cr are probably stimulatory to plant growth, particularly in soil where Mo is limiting. Both Cr(III) and Cr(VI) are redox- and pH-sensitive, with redox reactions converting them from one state to the other under oxic environments. Under anoxic conditions, the environment is too reducing to permit redox reactions. The interconversion of Cr from one state to the other depends on the presence of electron acceptors (i.e. oxidants such as O_2, Mn oxides) and electron donors (i.e. reductants such as organic matter, sulfides and Fe^{2+}. Its mobility in soil depends on the predominant species, with the hexavalent form being more mobile that the trivalent form (Baveye *et al.*, 1999).

$$e \text{ donor (anoxic)}$$
$$Cr(VI) \equiv Cr(III)$$
$$e \text{ acceptor (oxic)}$$

The fate of chromium in soils depends on the oxidation state, redox potential, pH, presence of electron acceptors and donors, competing ions, soil minerals, and organic matter content, which affects the partitioning of Cr between the solid and aqueous phases. Of these factors, prevailing soil pH seems to be the most important factor affecting Cr bioavailability. Thus, the adsorption of the toxic Cr(VI) in soils decreases with increasing pH, making it highly mobile above pH 8.5, whereas, adsorption of Cr(III) increases under alkaline conditions (Adriano, 2001). In other words, dissolution and bioavailability of the more toxic Cr(VI) is enhanced under alkaline conditions whereas, the availability of the less toxic Cr(III) is reduced under these same conditions. This observation has practical implications in systems where lime-treated biosolids are applied to agricultural fields. Liming of biosolids to pH 11-12 is routinely practiced in some sewage treatment systems as to effectively inactivate enteroviruses and helminths (US EPA, 1992; Jjemba, 2004). This practice may inadvertently enhance the bioavailability of some more toxic species of heavy metals such as soil Cr if the final soil pH is elevated to alkaline levels. This contention highlights the importance of considering how biosolids were treated prior to application to land and may, at least in part, explain why reports about the bioavailability of heavy metals after applying sludge give inconsistent results if factors such as the nature of biosolids used for the study are not elucidated.

Under normal soil conditions, the reduction of Cr(VI) to Cr(III) is favored due to the presence of reducing agents such as organic matter, Fe(II) and sulfides. Cr(III) is also susceptible to sorption by organic matter and Fe oxides (Adriano, 2001). Such sorption can minimize the oxidation of Cr(III) to Cr(VI) at Cr-polluted sites if livestock manure (Losi et al., 1994) or non-lime treated biosolids are applied as part of the remediation strategy. Thus, contrary to the scenario described above, sewage sludge application has, in some instances, been shown to reduce the bioavailability of heavy metals in polluted soils (Moreno et al., 2002).

Plant roots and exudates have a significant influence on the speciation of Cr and its subsequent uptake. Once inside the plant, chromium remains more concentrated in the roots due to some form of barrier effect of the cell wall to translocating this metal from roots to the shoot. Bioaccumulation levels of Cr in plants are typically less than 1 mg kg^{-1} and in general, concentrations of 1-5 mg kg^{-1} Cr(III) or Cr(VI) in bioavailable form in soil cause phytotoxicity to most plant species irrespective of the oxidation state (Adriano, 2001). Its mechanism of phytotoxicity is not clear but could be due to interference with Mo, Fe, P, and N metabolism. There are exceptions, however; accumulator plants such as *Sutera folina*, *Leptospermum scoparium*, *Geissois* sp., and *Dicoma miccolifera* can accumulate concentrations that are much higher than 5 mg kg^{-1}.

Arsenic

The concentration of As in the earth's crust is only 1.8 mg kg^{-1} but concentrations as high as 1059 mg kg^{-1} soil have been documented at contaminated sites (Greenwood and Earnshaw, 1997; Schnoor, 2000). Arsenic occurs as arsenite (III), arsenate (V) or their methylated derivatives, the dynamics between each state being achieved by both chemical and microbiological mechanisms (Dowdle *et al.*, 1996). Arsenite is more toxic than arsenate (V) which in turn is more toxic than the methylated forms of As. Both arsenite (III) and arsenate (V) are mainly encountered in soils whereas the methylated forms primarily occur in sediments and other anoxic environments (Jjemba, 2004). Humans are primarily exposed to As through inhalation and ingestion with the latter mostly occurring in drinking water. Uptake by plants is another route through which As enters the food chain. Fortunately, ingested As has a shorter half-life in the body than inhaled As due to a more rapid biotransformation in the liver (Adriano, 2001).

The chemical species of As influences its sorption affinity and its subsequent bioavailability. In general, As (V) has a higher affinity for soils and sediments than the more toxic As (III). This observation has implications in treating environmental contamination in that As-contaminated soils can be managed by creating (oxic) conditions that favor the oxidation of the contaminant to the benign and less mobile (i.e., less bioavailable) As (V). The chemistry of As(V) is similar to that of phosphorus, its immediate neighbor in the Periodic table. The mobility and bioavialability of both As(III) and As(V) is high in poorly drained soils (McBride, 1994).

The bulk concentration of arsenic in soils does not typically reflect the amount that is bioavailable. In a phytoremediation study reported by Schnoor (2000) at a very heavily As-contaminated site, there was only a weak correlation (R^2=0.3) between As in soil and As in the plant stem. Typical extractants for As from soil are distilled water, 0.1 N KNO_3, 0.05 N HCl (individually or as a mixture with 0.025 N H_2SO_4), 0.1 N $NaHCO_3$, or 0.05 M KH_2PO_4. Both the mixed acid and 0.5 N $NaHCO_3$ are also widely used for extraction and correlate well with plant growth (r=0.81 and 0.82, respectively) in predicting the bioavailable As (Adriano, 2001).

Cadmium

Similar to the other two group IIB transition metals (i.e., Zn and Hg), cadmium has only one oxidation state in its stable form (Table 2.2). The primary Cd sources to the environment include electroplating (auto industry), pigments production, the plastics industry, anti-corrosion of steel, galvanized metal production, fungicide production and disposal of rechargeable Ni-Cd batteries. Cadmium occurs in the earth's crust at concentrations of a meager 0.16 mg kg^{-1} (Greenwood and Earnshaw, 1997) although sites polluted with 22-350 mg kg^{-1} soil have been reported (Adriano, 2001). Typical concentrations of Cd in agricultural soils in the US range between <0.01-1.4 mg kg^{-1} soil (Holmgren et al., 1993).

The mobility of cadmium in soils is still a source of controversy. For example, McGrath and Lane (1989) measured low Cd mobility in soil. In contrast, Baveye et al. (1999) found an increased concentration to a 75 cm depth in a sludge-treated silt loam soil. It appears that pH, organic matter content, texture, as well as the form in which the ion occurs in the soil affect its sorption and mobility (Probst et al., 2003). Sequential extraction procedures have shown that Cd partitions into exchangeable, carbonate-bound, Fe-Mn oxides-bound, and organic-bound fractions (Adriano, 2001). Under acidic soil conditions, Cd ions exchange into the soil solution phase. Its bioavailabilty is also enhanced by the presence of other competing cations such as Zn, Cu, and Cr, possibly due to their relatively smaller ionic radii compared to Cd (Table 2.2). This argument is in agreement with Morera et al. (2001) who, using hydrolysis constants (pK values), found that Cu=Pb (pK=7.7) is less bioavailable than Zn (pK=9.0) which in turn is less bioavailable than Cd (pK=10.1). This observation is further supported by the large BSAF values of Cd in two earthworm species at a site that was contaminated by various heavy metals (Fig. 2.1). The tendency of Cd to experience such enhanced bioavailability in the presence of other metals has serious ramifications from a practical perspective, as heavy metal polluted sites typically contain a suite of metals. When present in the growth matrix at higher than background concentrations, Cd is readily taken up by existing biota. Plants are particularly known to take up Cd in a linear fashion; thus, where grown in soils with high Cd concentrations, high concentrations of the metal have also been reported in the vegetation (Shacklette, 1972 as cited by Adriano 2001).

Lead

Lead typically occurs naturally in soils at concentrations ranging between 1-200 mg kg^{-1} with a mean concentration of 13 mg kg^{-1} (Greenwood and Earnshaw, 1977). It occurs in oxidation states I to IV but only Pb(II) is relevant in the terrestrial environment. Its presence in terrestrial environments is a result of previous applications in lead-based paint, lead-based pesticides such as lead arsenate, plumbing and as an anti-knock additive in gasoline. While the use of lead-containing gasoline has been essentially phased out in most developed countries, its use in developing countries continues.

Lead retention in soils increases with higher organic carbon contents and cation exchange capacity, although calcite, oxides and hydroxides of Al, Fe, and Mn also serve as important sorbents of Pb (Chirenje *et al.*, 2004). On the other hand, Pb solubility and subsequent mobility as well as bioavailability are enhanced as the soil pH decreases (Suave *et al.*, 1998). As with arsenic, lead can be methylated through biological processes, the resulting methyllead and ethyllead inhibiting other important natural processes such as the transformation of nitrogen in soils and sediments (Jjemba, 2004).

As is the case with most heavy metals, the total content of Pb in soil does not adequately reflect its bioavailability in terrestrial environments, which markedly depends on pH, organic matter content, clay content, and type of biota. Lead mobility in the soil is quite limited (Elsokkary *et al.*, 1995). In fact, Pb is the least mobile heavy metal in soil, especially under reducing or alkaline conditions. It is also greatly immobilized by organic matter and oxides (e.g., Al_2O_3). However, substantial movement of Pb has been reported in a slightly acidic (pH 5.9) silty clay loam (Stevenson and Welch, 1979). Lead bioavailability is represented by the extractable amounts, which vary depending on the extract used. Common extractants are 0.05M EDTA, DTPA, HCl, 0.5M BaCl, HOAc, and NH_3HCO_3. Lead is less readily extracted at alkaline pH values due to its ability to chemisorb on oxides and clays or form highly stable complexes with phosphates, hydroxides and carbonates (McBride, 1994). Its bioavailability is also greatly diminished when phosphate fertilizers are applied to agricultural fields, due to formation of complexes (Hettiarachchi *et al.*, 2000).

From the soil, plants can passively accumulate Pb although the bioavailable fraction is usually low even in cases where total soil Pb content is high (Wu *et al.*, 1999). Furthermore, most of the accumulated metal is not readily translocated to the shoot and grain (McBride, 1994), an observation that has implications on the ultimate use for which crops grown in Pb-contaminated soil are intended.

CONCLUDING REMARKS

It is apparent from the above discussion that regulatory actions need to draw a distinction between metal species when establishing acceptable limits regarding application to soil. An effort to integrate recent advances in molecular techniques such as those of Ivask *et al.* (2004) using bioluminescence offer great potential in understanding the bioavailability and toxicity of metals in terrestrial systems. It is also important to realize that metal extraction procedures and extractants are quite diverse, and in most instances result in dramatic differences in extraction efficiency. A variety of plants are known to hyperaccumulate heavy metals and phytoremediation is increasingly being practiced as a remediation strategy. This is an area of current intensive research that, among other challenges, must confront the fact that even proficient hyperaccumulators are affected by metal speciation.

REFERENCES

Adriano, D.C. (2001). Trace Elements in Terrestrial Environments: Biogeochemistry, Bioavailability, and Risks of Metals. Springer-Verlag New York, Inc., New York, NY.

Athar, R., Ahmad, M. (2002). Heavy metal toxicity in legume-microsymbiont system. Journal of Plant Nutrition 25: 369-386.

Atkinson, B.W., Bux, F., Kasan, H.C. (1998). Waste activated sludge remediation of metal-plating effluents. Water SA 24: 355-359.

Babich, H., Stotzky, G. (1983). Synergism between nickel and copper in their toxicity to microbes: mediation by pH. Ecotoxicology and Environmental Safety 6: 577-589.

Baladane, M., Faivre, D., van Oort, H., Dahmani-Muller, H. (1999). Mutual effects of soil organic matter dynamics and heavy metals fate in a metallophyte grassland. Environmental Pollution 105: 45-54.

Baveye, P., McBride, M.B., Bouldin, D., Hinesly, T.D., Dahdoh, M.S.A., Abdel-Sabour, M.F. (1999). Mass balance and distribution of sludge-borne trace elements in a silt loam soil following long-term applications of sewage sludge. Science of Total Environment 227: 13-28.

Berti, W.R., Jacobs, L.W. (1996). Chemistry and phytotoxicity of soil trace elements from repeated sewage sludge applications. Journal of Environmental Quality 25: 1025-1032.

Bhogal, A., Nicholson, F.A., Chambers, B.J., Shepherd, M.A. (2003). Effects of sewage sludge additions on heavy metal availability in light textured soils: implications for crop yields and metal uptakes. Environmental Pollution 121: 413-423.

Chander, J., Brookes, P.C. (1991). Effect of heavy metals from past applications sewage sludge on microbial biomass and organic accumulation in a sandy and silty loam soil. Soil Biology and Biochemistry 23: 927-932

Chirenje, T., Ma, L.Q., Reeves, M., Szulcze, M. (2004). Lead distribution in near-surface soils of two Florida cities: Gainesville and Miami. Geoderma 119: 113-120.

Christie, P., Kilpatrick, D.J. (1992). Vesicular-arbuscular mycorrhiza infection in cut grassland following longterm slurry application. Soil Biology and Biochemistry 24: 325-330.

Cortet, J., Vauflery, A.G.D., Balaguer, N.P., Gomot, L., Texier, Ch., Cluzeau, D. (1999). The use of invertebrate soil fauna in monitoring pollutant effects. European Journal of Soil Biology 35: 115-134.

Dai, J., Becquer, T., Rouller, J.H., Reversat, G., Bernhard-Reversat, F., Nahmani, J., Lavelle, P. (2004). Heavy metal accumulation by two earthworm species and its relationship to total and DTPA-extractable metals in soils. Soil Biology and Biochemistry 36: 91-98.

Dowdle, P.R., Lavernman, A.M., Oremland, R.S. (1996). Bacterial dissimilatory reduction of arsenic (V) to arsenic (III) in anoxic sediments. Applied and Environmental Microbiology 62: 1664-1669.

Elsokkary, I.H., Ames, M.A., Shalaby, E.A. (1995). Assessment of inorganic lead species and total organo-alkyllead in some Egyptian agricultural soils. Environmental Pollution 87: 225-233.

Elzinga, E.J., van Grinsven, J.J.M., Swartjes, F.A. (1999). General purpose Freundlich isotherms for cadmium, copper and zinc in soils. European Journal of Soil Science 50: 139-149.

Federal Register (1990). Toxicity Characteristic Leaching Procedure (1990) 51: 21672-21692.

Giller, K.E., Witter, E., McGrath, S.P. (1998). Toxicity of heavy metals to microorganisms and microbial processes in agricultural soils: A review. Soil Biology and Biochemistry 30: 1389-1414.

Green, I., Merrington, G., Tibbett, M. (2003). Transfer of cadmium and zinc from sewage sludge amended soil through a plant-aphid system to newly emerged adult ladybirds (*Coccinella septempunctata*). Agriculture Ecosystem and Environment 99: 171-178.

Greenwood, N.N., Earnshaw, A. (1997). Chemistry of the Elements. 2nd Edition. Butterworth-Heinemann, Boston.

Haq, A.U., Bates, T.E., Soon, Y.K. (1980). Comparison of extractants for plant-available zinc, cadmium, nickel, and copper in contaminates soils. Soil Science Society of America Journal 44: 772-777.

Harvey, C.F., Swart, C.H., Badruzzaman, A.B.M., Keon-Blute, N., Yu, W., Ashraf, M.A., Jay, J., Beckie, R., Niedan, V., Brabander, D., Oates, P.M., Ashfaque, K.N., Islam, S., Hemond, H.F., Ahmed, M.F. (2002). Arsenic mobility and groundwater extraction in Bangladesh. Science 298: 1602-1606.

Hettiarachchi, G.M., Pierzynski, G.M., Ransom, M.D. (2000). In situ stabilization of soil lead using phosphorus and manganese oxide. Environmental Science and Technology 34: 4614-4619.

Holmgren, G.G.S., Meyer, M.W., Chaney, R.L., Daniels, R.B. (1993). Cadmium, lead, zinc, copper and nickel in agricultural soils of the United States of America. Journal of Environmental Quality 22: 335-348.

Horswell, J., Speir, T.W., von Schaik, A.P. (2003). Bio-indicators to assess impacts of heavy metals in land-applied sewage sludge. Soil Biology and Biochemistry 35: 1501-1505.

Insam, H., Hutchinson, T.C., Reber, H.H. (1996). Effects of heavy metal stress on the metabolic quotient of the soil microflora. Soil Biology and Biochemistry 28: 691-694.

Ivask, A., François, M., Kahni, A., Dubourguier, H-C., Virta, M., Dousy, F. (2004). Recombinant luminescent bacterial sensors for the measurement of bioavailability of cadmium and lead in soils polluted by metal smelters. Chemosphere 55: 147-150.

Jjemba, P.K. (2004). Environmental Microbiology: Principles and Applications. Science Publishers, Inc., Enfield New Hampshire.

Khan, M., Scullion, J. (2002). Effects of metal (Cd, Cu, Ni, Pb or Zn) enrichment of sewage-sludge on soil micro-organisms and their activities. Applied Soil Ecology 20: 145-155.

Kuperman, R.G., Carrierro, M.M. (1997). Soil heavy metal concentrations, microbial biomass and enzyme activities in a contaminated grassland ecosystem. Soil Biology and Biochemistry 29: 179-190.

Lai, H-Y., Chen, Z-S. (2004). Effects of EDTA on solubility of cadmium, zinc, and lead and their uptake by rainbow pink and vetiver grass. Chemosphere 55: 421-430.

Logan, T.J., Lindsay, B.J., Goins, L.E., Tyan, J.A. (1997). Field assessment of sludge metal availability to crops: sludge rate response. Journal of Environmental Quality 26: 534-550.

Losi, M.E., Amthein, C., Frankenberger, W.T. (1994). Factors affecting chemical and biological reduction of hexavalent chromium in soil. Journal of Environmental Quality 23: 1141-1150.

Manios, T., Stentiford, E.I., Millner, P. (2003). Removal of heavy metals from a metaliferous water solution by *Typha latifolia* plants and sewage sludge compost. Chemosphere 53: 487-494.

Matthews, P. (1996). A Global Atlas of Wastewater Sludge and Biosolids Use and Disposal. International Association on Water Quality, London.

McBride, M.B. (1994). Environmental Chemistry of Soils. Oxford University Press, New York.

McGrath, S.P., Cegarra, J. (1992). Chemical extractability of heavy metals during and after long-term applications of sewage-sludge to soil. Journal of Soil Science 43: 313-321.

McGrath, S.P., Lane, P.W. (1989). An explanation for the apparent losses of metals in a long-term field experiment with sewage sludge. Environmental Pollution 60: 235-256.

Mercier, G., Duchesne, J., Carles-Gibergues, A. (2002). A simple and fast screening test to detect soils polluted by lead. Environmental Pollution 118: 285-296.

Moreno, J.L., Hernández, T., Pèrez, A., García, C. (2002). Toxicity of cadmium to soil microbial activity: effect of sewage sludge addition to soil on the ecological dose. Applied Soil Ecology 21: 149-158.

Morera, M.T., Echeverria, J.C., Mazkiaran, C., Garrido, J.J. (2001). Isotherms and sequential extraction procedures for evaluating sorption and distribution of heavy metals in soils. Environmental Pollution 113: 135-144.

Naseby, D.C., Lynch, J.M.(1997). Rhizosphere soil enzymes as indicators of perturbations caused by enzyme substrate addition and inoculation of genetically modified strain. Soil Biology and Biochemistry 29: 1353-1362.

Newman, M.C., Jagoe, C.H. (1994). Ligands and the bioavailability of metals in aquatic environments. In: Bioavailability: Physical, Chemical, and Biological Interactions. pp. 39-61. Eds. J.L. Hamelink, P.F. Landrum, H.L. Bergman and W.H. Benson. Lewis Publishers, Boca Raton.

Oliver, M.A. (1997). Soil and human health: a review. European Journal of Soil Science 48: 573-592.

Pascaul, J.A., Garcia, C., Hernandez, T. (1999). Casting microbiological and biochemical effects of the addition of municipal solid waste to an arid soil. Biology and Fertility of Soils 30: 1-6.

Probst, A., Hernandez, L., Probst, J.L. (2003). Heavy metal partitioning in three French forest soils by sequential extraction procedure. Journal de Physique IV French 107: 1103-1106.

Poon, C.S., Lio, K.W. (1997). The limitation of toxicity characteristic leaching procedure for evaluating cement-based stabilised waste forms. Waste Management 17: 15-23.

Quevauviller, P., Rauret, G., Lopez-Sanchez, J.-F., Rubio, R., Ure, A., Mantua, H. (1997). Certification of trace metal extractable contents in a sediment reference material (CRM 601) following a three-step sequential extraction procedure. Science of the Total Environment 205: 223-234.

Renella, G., Ortigaza, A.L.R., Landi, L., Nannipieri, P. (2003). Additive effects of copper and zinc on a cadmium toxicity on phosphatase activities at ATP content of soil as estimated by the ecological dose (ED_{50}). Soil Biology and Biochemistry 35: 1203-1210.

Rost, U., Joergensen, R.G., Chanders, K. (2001). Effect of Zn enriched sewage-sludge on microbial activities and biomass in soil. Soil Biology and Biochemistry 33: 633-638.

Schnoor, J.L. (2000). Phytostabilization of metals using hybrid poplar trees. In: Phytoremediation of Toxic Metals: Using Plants to Clean up the Environment. pp. 133-150. Eds. I. Raskin and B.D. Ensley. John Wiley and Sons, Inc., New York.

Smith, S.R. (1997). *Rhizobium* in soils contaminated with copper and zinc following the long-term application of sewage sludge and other organic wastes. Soil Biology and Biochemistry 29: 1475-1489.

Snyman, H.G. (2001). Characterization of sewage sludge metals for classification purposes using the potentially leacheable metal fraction. Water Science and Technology 44: 107-114.

Stehouwer, R.C., Wolf, A.M., Doty, W.T. (2000). Chemical monitoring of sewage sludge in Pennsylvania: variability and application uncertainty. Journal of Environmental Quality 29: 1686-1695.

Stevenson, F.J., Welch, L.F. (1979). Migration of applied lead in a field soil. Environmental Science and Technology 26: 966-974.

Suave, S., McBride, M., Hendershot, W. (1998). Soil solution speciation of Pb(II): effects of organic matter and pH. Soil Science Society of America Journal 62: 618-621.

Tessier, A., Campbell, P.C.G., Bisson, M. (1979). Sequential extraction procedure for the speciation of particulate trace metals. Analytical Chemistry 51: 844-851.

US EPA (1992). Environmental Regulations and Technology. Control of Pathogens and Vector Attraction in Sewage Sludge. Report No. EPA/625/R-92/013. Office of Research and Development, Washington D.C.

Wu, J., Hsu, F.C., Cunningham, S.D. (1999). Chelate-assisted Pb phytoextraction: Pb availability, uptake, and translocation constraints. Environmental Science and Technology 33: 1898-1904.

CHAPTER 3

Microbially Mediated Changes in the Mobility of Contaminant Metals in Soils and Sediments

Flynn Picardal[1] and D. Craig Cooper[2]
[1]*Indiana University, Bloomington, IN, 47405 USA.*
[2]*Idaho National Laboratory, Idaho Falls, ID, 83415 USA.*

Contamination of soils and deeper sediments by toxic metals, metalloids, and radionuclides is a worldwide problem due to improper disposal practices, spills, atmospheric deposition of combustion emissions, and intentional application of sludges, fertilizers, or other materials. Such soil contaminants can pose various ecological problems, leach into drinking and agricultural water sources, or enter the food chain via plant uptake and bioaccumulation in plants or animals. The potential for causing ecological damage or negative human health consequences is largely a function of metal redox state, solubility, speciation, and other physicochemical properties. Although many of these contaminants are strongly sorbed by secondary clay minerals, metal oxides, carbonates, and humic substances in soils, their mobility can be strongly affected by microbial activity by a variety of mechanisms.

Changes in metal mobility can result from direct microbial oxidation or reduction of the metal, direct sorption by microbial biomass, uptake and bioaccumulation of metals, indirect microbially mediated changes in metal redox state, indirect stimulation of mineral precipitation by microbial surfaces, and/or indirect changes in groundwater and sediment geochemistry as a result of microbial metabolic processes. Microbial reduction of chromate, for example, may convert highly mobile Cr(VI) to less mobile Cr(III). In other cases, changes in metal redox state may be indirectly caused by microbial activity. Cr(VI), for example, may also be indirectly reduced to Cr(III) by Fe(II)-containing minerals formed by the activity of iron reducing bacteria. Microbial metabolism can also produce complexing ligands and carbonate species that may alter aqueous metal speciation or solubility, and release or immobilize metals as a result of dissolution and precipitation events. All of these processes may be accelerated by the presence of humic compounds which may both serve as metal complexants and electron shuttles between the bacterial respiratory system and the mineral surface. Considered together, these various mechanisms present a variety of options for the microbially mediated mobility of contaminant metals in soils and sediments.

INTRODUCTION

Contamination of soils by heavy metals arises from a large number of activities including unintentional spills or releases from industrial processes, mining and metal refining operations, volatilization during incineration or combustion processes and subsequent atmospheric deposition, soil application of metal-containing sludge biosolids or flyash, land treatment of oily wastes, application of fertilizers and pesticides, and vehicle emissions. Remediation of metal-contaminated soils is broadly limited to three approaches. Firstly, the soil can be excavated and either treated or transported to a secure landfill or disposal facility. Secondly, the soil can be treated in place in such a way that the contaminant metals are less mobile, less bioavailable, or less toxic. Lastly, the soil can be left in place but contained so that the metals are not available to plants or animals and not able to move away from the contained site.

Treatment options are obviously more limited for metallic contaminants than for organic contaminants since metals cannot be destroyed. Methods used to treat metal contamination in soils (Peters and Shem, 1995) can include removal by a soil washing process that separates the finer soil particles which frequently contain a greater weight percentage of the metal contaminant. This can be followed by extractions using acidic solutions, chelating agents, oxidizing / reducing agents, or other chemical treatments to render the metal more extractable. Alternately, the contaminated soil might be treated in place by adding amendments that would immobilize the metals in a non-leachable, cement-like matrix.

Bioremediation of metal-contaminated soils also focuses on *in situ* treatment but uses microbial activity to mitigate harmful potentials by making the metals less mobile, bioavailable, or toxic. Bioremediation can be accomplished using several approaches. Biostimulation involves the addition of nutrients to stimulate the growth or activity of existing microorganisms that are able to carry out the desired metal transformation. Bioaugmentation, on the other hand, involves the addition of novel microorganisms that are able to transform the contaminants to a greater extent or at faster rates than the existing soil biota. Regardless of whether bioremediation is accomplished using bioaugmentation or biostimulation, the process relies on microbial alterations of metal speciation that alter metal mobility, availability, or toxicity. Although these three characteristics are in many cases related, our emphasis in this chapter will be on the first, i.e., microbially-mediated changes in contaminant metal speciation and mobility.

Understanding the speciation, fate, and transport of heavy metals in soils would be much easier if the topic could be approached solely as a static, chemical problem, i.e., one that could be accurately approximated through equilibrium-based, speciation models using the vast thermodynamic database that has been tabulated over the years. Clearly, however, microbiological processes result in non-equilibrium conditions that are often quite divergent from those predicted using thermodynamic models. Over the last 25 years, knowledge of microbial interactions with metals in soils has grown geometrically, and the interdisciplinary nature of this field requires some understanding of solid- and aqueous-phase geochemistry, metal toxicity, microbial metabolism and metal uptake, and microbial ecology. Together with new spectroscopic methods that can yield information about microbe-solid interfaces at the molecular level, e.g. X-ray absorbance spectroscopy, we are now beginning to gain an understanding of some of the exceedingly complex and dynamic biogeochemical interactions occurring in soils.

Metals are present in many different forms in soils, and the form(s) of the metal will determine its mobility and bioavailability. Although some metals form highly soluble species, e.g., the oxyanions of Se, As, and Cr(VI), most trace metals in soil are found associated with the soil matrix. In most cases, the total aqueous concentration (including free ions and soluble complexes) represent only 0.001% to 20% of the total soil metal content, the wide range dependent on the particular metal, soil composition, and conditions (Sauvé, 2002). The majority of the trace heavy metals are primarily associated with a number of soil components such as natural organic matter (NOM), Fe- and Mn-(hydr)oxides, silicate clays, carbonates, sulfides, residual parent material, and microbial biomass, although one particular soil component may control sorption of particular metals in specific soils. Xing *et al.*, for example, studied sorption and retention of heavy metals in two different soils from China and determined that the retention of Cr, Ni, Cu, Zn, and Co were primarily controlled by amorphous Fe (hydr)oxides, whereas Pb and V retention was chiefly controlled by sorption on layered aluminosilicate clays (Xing *et al.*, 1995). In general, metal speciation, mobility, and bioavailability in soils are ultimately determined by processes or conditions such as adsorption to mineral and microbial surfaces, formation of complexes with aqueous or solid-phase NOM, soil solution saturation and metal precipitation or dissolution, redox processes, pH, composition of the soil matrix, water potential, and many other physical, chemical, or biological processes.

Interactions between heavy metals and microbes in soils and sediments include both the effects of metals on microorganisms as a result of nutritional requirements and toxicity, and the influence of microbial activity on the speciation and mobility of the metals. Microbial uptake of trace metals, metal assimilation, and metal toxicity have been widely studied (for example, see Sunda and Huntsman, 1998; Jentschke and Godbold, 2000; Blencowe and

Morby, 2003; Mergeay *et al.*, 2003; Mulrooney and Hausinger, 2003) and this review will focus only on the influence of microbial activity on metals speciation and mobility. As a result of research funding from the US Department of Energy (Atkinson *et al.*, 2003) and other agencies concerned with the environmental consequences of nuclear power generation and weapons production, there is a significant body of literature that deals with microbial interactions with metallic radionuclides, e.g., U (Lovley *et al.*, 1991; Senko *et al.*, 2002; Shelobolina *et al.*, 2004), Tc (Lloyd *et al.*, 2000a; Wildung *et al.*, 2000), or Np (Lloyd *et al.*, 2000b). Indeed, the contamination of soils and deeper sediments with metals and radionuclides is a major environmental problem at many US government facilities (Riley *et al.*, 1992) and much work has been done to develop bioremediation methods that would be effective in immobilizing or detoxifying some of these pollutants (Atkinson *et al.*, 2003). The present review, however, will focus on the non-radioactive heavy metals and metalloids, e.g., Pb, As, Zn, Cr, Cu, Cd, Ni, Co, etc., that are a concern in contaminated soils at a larger number of sites and more widespread than radionuclides.

There are numerous ways in which microbial activity can effect mobility of contaminant metals. Firstly, *direct* microbial reduction or oxidation of the metal can cause changes in metal speciation, sorptive properties, and solubility. Similar changes in the oxidation state of the metal might occur *indirectly* as a result of microbial production of redox reactive species, e.g., $Fe(II)$, that may then chemically reduce the metal. Additionally, previously surface-bound and bioavailable metals may become immobilized and unavailable to biota if they are incorporated into new, stable minerals formed as a result of microbially catalyzed changes in soil mineralogy. Microbially-mediated changes in soil solution, pH, and production of siderophores and other organic complexants may also have a dramatic effect on mobility. Lastly, the microorganisms themselves can sorb metals and serve as a nucleation site for metal precipitates. Although fungi may play an important role in microbe-metal interactions in soils (see Gadd and Sayer, 2000, for a review), our focus in this review will be on (i) direct and indirect bacterial reduction and oxidation of metals and (ii) the geochemical consequences of metal sorption on bacteria.

DIRECT MICROBIAL REDUCTION AND OXIDATION OF METALS AND EFFECTS ON METAL MOBILITY

Microbial Redox Transformations of Iron

Although microorganisms can metabolize metals through biosynthetic pathways, we will primarily be concerned with reactions in which bacteria oxidize or reduce metals in the soil or sediment matrix and thereby affect metal toxicity or mobility. Since iron is the most common, redox-reactive metal in soils due to the frequent presence of iron (hydr)oxides and iron-bearing clays, the use of Fe(III) in microbial energy generation arguably represents the most significant environmental example of microbial metabolism of metals. Although some fermentative bacteria are able to reduce Fe(III) during anaerobic growth (see Lovley, 2000 for a review), in such cases the amount of Fe(III) reduced is relatively small and Fe(III) is not required for growth. In dissimilatory Fe(III) reduction (also known as Fe(III) respiration), Fe(III) can be reduced relatively rapidly, in large amounts, and can serve as the electron acceptor for microbial growth in anoxic soils and sediments. In this respiratory process, an electron donor (hydrogen or an organic compound such as acetate, formate, lactate, etc.) is oxidized at the expense of the terminal electron acceptor, Fe(III), which is reduced to Fe(II). The energy from the electrons transferred in this redox reaction is used to generate a proton gradient across the bacterial membrane and, ultimately, used to phosphorylate ATP.

Iron is the fourth most abundant element in the earth's crust and forms stable minerals in both the divalent (e.g., siderite, vivianite, ferrous sulfides) and trivalent (e.g., goethite, lepidocrocite, hematite) oxidation states (Cornell and Schwertmann, 1996). The biogeochemical cycling of iron has been the subject of much study over the last two decades and it has become clear that Fe(III) minerals can be the dominant electron acceptor for oxidation of organic matter in anoxic, terrestrial and non-marine sediments. As a result of large, specific surface areas and reactive surface hydroxyl groups, iron (hydr)oxides are a major sorbent of contaminant metals in soils and sediments, and microbially mediated changes in the sorbent can have a dramatic effect on the speciation of the sorbed or coprecipitated metal (Francis and Dodge, 1990; Cooper *et al.*, 2000). Since Fe(II) is a strong reductant able to abiotically transform a wide number of organic (Heijman *et al.*, 1993; Kim and Picardal, 1999) and inorganic compounds (discussed below), microbial Fe(III) reduction and iron redox chemistry can exert a strong influence on overall soil geochemistry and the fate of soil pollutants.

Many Fe(III) (hydr)oxides and Mn(IV) oxides can serve as a respiratory oxidant under anoxic conditions for numerous dissimilatory iron- and manganese-reducing bacteria (DIRBs and DMRBs) (Nealson and Myers, 1992; Lovley, 1993; Nealson and Saffarini, 1994; Lovley, 2000). Since many bacteria are capable of both Fe(III) and Mn(IV) reduction, the two processes are often considered together. In our review, we will cover only Fe(III) reduction and the reader is referred to recent, excellent reviews for additional information on microbial Mn redox transformations (Ehrlich, 1996b; Emerson, 2000; Lovley, 2000).

Microbial iron reduction is inhibited by exposure to oxygen and is an anaerobic process that can take place in waterlogged soils or at anaerobic microsites in otherwise oxic soils. The extent of water-filled porosity and concentration of oxygen-consuming organic matter will therefore, to a large extent, determine if significant rates of iron reduction are possible, even if DIRBs and reducible Fe(III) (hydr)oxides are present. Although many DIRBs are strict anaerobes, e.g., *Geobacter* sp., whose activity is expected to be limited to anoxic zones, other DIRBs, such as *Shewanella putrefaciens*, are facultative anaerobes. The latter type of microorganism may have a competitive advantage in soils that undergo periodic redox fluctuations as a result of intermittent infiltration of water and/or organic matter.

Microbial Fe(III) reduction can produce a suite of changes in the mineralogy of the soil inorganic fraction which may, as we will discuss below, have a notable effect on the mobility of metals sorbed to soil solids. A simple reaction showing the oxidation of acetate to bicarbonate coupled to the reduction of Fe(III) in goethite (α-FeOOH) to aqueous Fe^{2+} is shown in equation (1). Redox reactions showing Fe(III) reduction can be written in various stoichiometries, depending not only on the carbon source, but also on the form of Fe(III) and Fe(II) involved. Instead of producing Fe^{2+} as in Equation 1, for example, a similar reaction could be written in which either magnetite (Fe_3O_4) or siderite ($FeCO_3$) was produced as an Fe(II)-containing product.

(1) $1/8\ CH_3COO^- + \text{-FeOOH} + 1\ 7/8\ H^+ \rightarrow Fe^{2+} + \frac{1}{4}\ HCO_3^- + 1\frac{1}{2}\ H_2O$

Fe(III) is relatively insoluble at circumneutral pH and typically will be present in soils and sediments as a wide range of amorphous to crystalline (hydr)oxides and Fe(III)-bearing clays. These materials will generally exist as coatings on other minerals in the soil and may be intermixed with natural organic matter (NOM). As these Fe(III)-containing minerals are reduced by DIRBs, biogenic Fe(II)-containing mineral phases are formed. Microbial reduction of lepidocrocite, for example, has produced magnetite in laboratory experiments (Cooper *et al.*, 2000). Magnetite is frequently observed during reduction of amorphous, hydrous ferric oxide (HFO) (Lovley *et al.*, 1987). In addition to resulting in mineralogical changes in iron minerals, microbial iron reduction has been shown to influence the swelling of iron-bearing clays (Kostka *et al.*, 1999b). Aqueous Fe^{2+} produced during mineral reduction can be transported with the soil solution to deeper horizons where it could sorb to solids or later reprecipitate as an Fe(III) (hydr)oxide if oxic conditions were established.

Most respiratory oxidants used by soil microorganisms are soluble, e.g., O_2 and NO_3^-. The mechanisms by which the solid and sparingly soluble Fe(III) oxides or iron-bearing clays are relatively rapidly reduced by soil microorganisms are still subject to much discussion and may be variable depending on species and conditions. Since contaminant heavy metals in a soil may be present as precipitates or sorbed to sesquioxides or clays, microbial metabolism of heavy metals may also depend on the presence of a mechanism to effectively access an extracellular, solid surface. For this reason, it is worthwhile to briefly review some of the proposed mechanisms that microorganisms may use to reduce solid-phase minerals in soils. Although the functional arrangement of respiratory electron- and hydrogen-carriers in the membranes of DIRBs remains to be elucidated, numerous cytochromes (Picardal *et al.*, 1993; Tsapin *et al.*, 2001), quinones (Myers and Myers, 1993) and Fe(III) reductases (Magnuson *et al.*, 2000; Kaufmann and Lovley, 2001) have been identified in the membranes of various *Shewanella* and *Geobacter* species. Some of these components are located in the outer membrane as opposed to most respiratory biochemicals which are located in the cytoplasmic membrane or periplasm (Myers and Myers, 1992; Myers and Myers, 1997; Beliaev *et al.*, 2001).

If an Fe(III) reductase were located on the outer membrane, it would solve the problem of electron transfer to a solid surface if the bacterium was able to make direct contact with the solid. Indeed, some studies in which the bacterium and oxide have been separated using semi-permeable membranes (Arnold *et al.*, 1988) or (alginate) beads (Nevin and Lovley, 2000) have shown that direct contact is required. The use of semi-permeable membranes in this context has been questioned (Nevin and Lovley, 2000), however, and direct contact does not always appear necessary (Nevin and Lovley, 2002).

Numerous studies have also shown that the rates of Fe(III) reduction can be increased by the presence of soluble electron shuttles. If the shuttle compound contained a moiety, e.g., a quinone, able to be alternately reduced by the microorganism and reoxidized by the solid oxide, it could serve as a reusable electron shuttle. Some DIRBs have been shown to produce such compounds (Newman and Kolter, 2000; Turick *et al.*, 2002) and recent studies using atomic force microscopy have shown that a crystalline oxide is reduced at locations other than where bacteria are attached, possibly suggesting the presence of a diffusible shuttle capable of carrying reducing electrons (Rosso *et al.*, 2003). Many DIRBs are also able reduce and utilize soil NOM as a respiratory electron acceptor, likely due to the presence of quinone-like moieties (Lovley *et al.*, 1996; Lovley and Fraga, 1998; Scott *et al.*, 1998). The reduced NOM is able to be reoxidized by Fe(III) (hydr)oxides and NOM has been shown to increase iron reduction rates, presumably by this electron-shuttling mechanism. Although direct contact and electron-shuttle models have both been presented to explain results in the laboratory, much work remains to be done to understand which mechanisms dominate reduction of solid minerals in soils.

In addition to serving as a terminal electron acceptor for microbial respiration when present as oxidized Fe(III), iron can serve as an electron donor for microbial energy generation when present as reduced Fe(II). The ability of acidophilic, lithotrophic bacteria, e.g., *Thiobacillus ferrooxidans*, to oxidize Fe(II) at low pH has been well-known for many decades (Brock and Gustafson, 1976; Harrison Jr., 1984). Our focus here, however, will be on bacteria able to oxidize Fe(II) at circumneutral pH.

Microbial oxidation of Fe(II) under aerobic conditions can be considered problematic due to the relatively rapid, abiotic oxidation of Fe(II) by molecular oxygen at circumneutral pH (Davison and Seed, 1983). Since Fe(II)-oxidizing bacteria (FeOB) must compete with this abiotic reaction, they are found primarily in environments where the chemical oxidation rate is minimized. This is clearly the case for the acidophilic FeOB since chemical Fe(II) oxidation by O_2 at low pH is slow. At circumneutral pH, abiotic oxidation rates will also be reduced at low O_2 concentrations, and growth of FeOB is commonly reported at redox interfaces or under microaerobic conditions (Kucera and Wolfe, 1957; Ehrlich, 1996a; Sobolev and Roden, 2002). As a result of the need for growth at redox gradients and the production of iron (hydr)oxides encrustations that can cover the cells, FeOB have developed a reputation as being difficult to obtain in pure culture (Hanert, 2004a). Nevertheless, there has been an increased interest in these microorganisms over the last 10 years and the development of a gel-stabilized, gradient enrichment technique by Emerson and Moyer, has resulted in the recent enrichment of several chemoautotrophs from diverse environments (Emerson and Moyer, 1997; Sobolev and Roden, 2004). Previously, the only FeOB thought to be capable of autotrophic growth was the stalk-forming *Gallionella ferruginea* (Hallbeck and Pederson, 1991; Hallbeck *et al.*, 1993; Hanert, 2004a).

Although lithoautotrophic FeOB may predominate in microaerobic environments with low organic carbon concentrations, many of the more well-known 'iron bacteria' are heterotrophs or mixotrophs, and accumulate iron (hydr)oxides on their surfaces for reasons that are not entirely understood (Emerson, 2000). Some of the best examples of such heterotrophic bacteria are the filamentous *Leptothrix* sp. and *Sphaerotilus* sp. (Spring, 2004) and the unicellular *Siderocapsa* and other related bacteria (Hanert, 2004b).

Over the last decade, it has also become evident that Fe(II) can be oxidized under anaerobic conditions. The first report of anaerobic oxidation of Fe(II) involved phototrophic bacteria isolated from freshwater and marine sediments that were able to use Fe(II) as an electron donor for anoxygenic photosynthesis (Widdel *et al.*, 1993; Ehrenreich and Widdel, 1994; Kappler and Newman, 2004). Although the significance of Fe(II) oxidation in soils by anoxygenic phototrophs is unknown, it was recently demonstrated that Fe(II) can also be utilized as a source of electrons for anaerobic nitrate reduction. Since nitrate-dependent Fe(II) oxidation was first discovered by Straub *et al.* in 1996, the enrichment of such bacteria from a wide variety of sources suggests that many facultative denitrifiers are able to oxidize both sorbed and soluble Fe(II) under anaerobic conditions (Straub *et al.*, 1996; Benz *et al.*, 1998; Straub and Buchholz-Cleven, 1998; Weber *et al.*, 2001). Most of the bacteria capable of nitrate-dependent Fe(II) oxidation apparently require an organic compound, e.g., acetate, at low concentrations at a carbon source (Straub *et al.*, 1996; Straub and Buchholz-Cleven, 1998). The existence of such bacteria able to anaerobically oxidize Fe(II) demonstrates the potential for the redox cycling of iron under anaerobic conditions in anoxic microsites or saturated soils.

Microbial Transformations of Chromium

Chromium exists as a soil contaminant primarily in the +3 and +6 oxidation states. Hexavalent Cr is much more toxic, at least in part due to the high solubility of the oxyanions, chromate (CrO_4^{2-}) and dichromate ($Cr_2O_7^{2-}$). These forms of Cr are also the species primarily present in industrial waste as a result of the widespread use of Cr in metal finishing, alloy production, and various other industrial processes. At circumneutral pH, trivalent Cr tends to precipitate as $Cr(OH)_3$, sorb to solid surfaces (Richard and Bourg, 1991), and is much less toxic. Microbial transformations of Cr that reduce Cr(VI) to Cr(III) therefore have the clear potential to reduce both Cr mobility and toxicity in soils.

Bacteria able to reduce Cr(VI) are widespread and, as opposed to many other metal-reduction processes, Cr(VI) reduction has been shown to occur under both oxic (Bopp and Ehrlich, 1988; Wang and Xiao, 1995; Garbisu *et al.*, 1998; McLean and Beveridge, 2001) and anoxic conditions (Ohtake *et al.*, 1990; Fude *et al.*, 1994; Lovley and Phillips, 1994; McLean and Beveridge, 2001). The ability to reduce Cr(VI) appears to be spread over diverse genera including species of *Pseudomonas* (Bopp and Ehrlich, 1988; McLean and Beveridge, 2001), *Shewanella* (Liu *et al.*, 2002; Viamajala *et al.*, 2003), *Enterobacter* (Ohtake *et al.*, 1990), *Desulfovibrio* (Lovley and Phillips, 1994), *Bacillus* (Garbisu *et al.*, 1998), and others (Wang and Shen, 1995). The widespread ability to reduce Cr(VI) under oxic and anoxic conditions suggests that different biochemical mechanisms may be utilized by different microorganisms. As described in a recent review by Wang (2000), aerobic Cr(VI) reduction primarily occurs via a soluble reductase, whereas anaerobic reduction may involve either or both a soluble or membrane-bound reductase. Although electron transport to Cr(VI) may not yield sufficient energy to allow growth of most Cr(VI) reducers (Lovley and Phillips; 1994, Wang, 2000), a strain of *Desulfotomaculum reducens* has been described that is able to utilize Cr(VI) and other metals as physiological electron acceptors for growth under anaerobic conditions (Tebo and Obraztsova, 1998).

Although Cr(VI)-reducing ability is not uncommon, laboratory research suggests the possibility that Cr(VI) reduction may be limited in some soils by toxic co-contaminants. Cr(VI) may itself be inhibitory at high concentrations. Schmieman and coworkers found that Cr(VI) concentrations of 12 mg l^{-1} inhibited growth of a Cr(VI)-reducing enrichment culture (Schmieman et al., 1997). In addition, some Cr(VI)-reducing bacteria have shown a finite reduction capacity in batch culture studies (Wang, 2000). This may be a result of Cr toxicity which ultimately terminates further Cr(VI) reduction. *Pseudomonas stutzeri* strains isolated from foundry soils, however, were able tolerate much higher Cr(VI) concentrations (Badar et al., 2000), presenting the possibility that long-term exposure to high concentrations in some soils may allow development of more resistant strains. Other metals, e.g., Zn^{2+} and Cu^{2+}, and phenolic compounds, e.g., phenol, *p*-cresol, and 2-chlorophenol, have also been shown to inhibit Cr(VI) reduction in batch culture (Wang and Shen, 1995). It is not clear if the toxicity shown in laboratory batch studies is directly applicable to the soil environment where sorption and other complex biological and geochemical interactions will occur. Other studies using soil columns pretreated with very high (1000 and 10,000 mg l^{-1}) concentrations of Cr(VI) showed that addition of organic carbon (lactate or tryptic soy broth) increased Cr(VI) reduction, although not all of the removal could be explained by direct enzymatic reduction (Tokunaga et al., 2003). Addition of organic carbon to Cr(VI)-contaminated soil in other studies similarly demonstrated increased amounts of Cr(VI) reduction (Turick et al., 1998).

Microbial Transformations of Arsenic and Selenium

As and Se are two toxic elements that are found at elevated and hazardous concentrations in soils both from natural and anthropogenic sources. Mobilization of Se by infiltration of irrigation water in the San Joaquin Valley, CA, USA, has contributed to toxic concentrations at the well-known Kesterson Reservoir with associated negative effects on wildlife (Ohlendorf et al., 1986). Release of As from alluvial aquifers in the Ganges delta has also resulted in contamination of irrigation and drinking water in Bangladesh and West Bengal (Smith et al., 2000). These elements have also found their way into soils from a variety of mining, industrial, and agricultural activities. Although As occurs in a number of oxidation states, the two most prevalent forms found in nature are the oxyanions arsenate (As(V); AsO_4^{3-}) and the partially reduced, more toxic arsenite (As(III); AsO_3^{3-} and AsO_2^{-}). Se also can be found in a number of valence states but the soluble oxyanions, selenate (Se(VI); SeO_4^{2-}) and partially reduced selenite (Se(IV); SeO_3^{2-}) are the primary forms found in oxic, seleniferous soils. In anoxic environments, insoluble elemental Se^0 is the dominant species.

Microorganisms, including both bacteria and fungi, participate in a number of reactions involved in cycling and (im)mobilization of Se and/or As, including assimilatory and dissimilatory redox transformations, methylation and demethylation, and volatilization. Here we will provide a concise description of some of the energy-producing bacterial redox transformations. The reader is referred to recent reviews (Dungan and Frankenberger, 1999; Stolz and Oremland, 1999; Oremland and Stolz, 2000) for additional information and a discussion of methylation and volatilization reactions.

It has been known for or 75 years that some bacteria able to reduce arsenate to arsenite (Green, 1918). Anaerobic reduction of As(V) to As(III) can be coupled to organic carbon oxidation by a reaction of the type:

(2) $lactate^- + 2\ HAsO_4^{2-} + 2\ H^+ \rightarrow 2\ H_2AsO_3^- + acetate^- + HCO_3^-$

Bacteria that respire As(V) are diverse, not constrained to any particular physiological group, and include species of *Bacillus* (Blum *et al.*, 1998), *Desulfotomaculum* (Newman *et al.*, 1997), and *Sulfurospirillum* (formerly *Geospirillum*) (Stolz *et al.*, 1999). In addition to organotrophic As(V) reduction, an autotrophic bacterium was recently isolated from lake sediments that is able to grow by coupling $S^=$-oxidation (to $SO_4^=$) to reduction of As(V) to As(III) (Hoeft *et al.*, 2004).

It is difficult to generalize about the overall consequences of As(V) reduction in soils because As(III) and As(V) are both toxic, soluble, and sorb to soil and sediment solids, the degree of sorption varying in accordance with geochemical characteristics of the soil. In general, however, As(III) is more toxic (Rosen, 2002) and mobile (Pierce and Moore, 1982) than As(V). If sorbed onto Fe(III)- or Mn(IV) (hydr)oxides, As can also be mobilized by reduction of the sorbant. Although *S. alga* BrY is unable to enzymatically reduce As, the bacterium was able to cause release of As(V) from a synthetic ferric arsenate (scorodite) and from sorption sites in natural sediments during reduction of Fe(III) (Cummings, 1999). As(III) was also released in other studies (Zobrist *et al.*, 2000) in which As(V) had been co-precipitated or sorbed to ferrihydrite and subsequently incubated with *Sulfurospirillum barnesii*, a bacterium capable of both Fe(III) and As(V) reduction. In those experiments, the rate of As(V) reduction was affected by how As was associated with the mineral phases present. Although the extent of As release is difficult to predict due to solid- and aqueous-phase geochemical considerations, such experiments clearly show that As(V) and Fe(III) reduction can play a significant role in the mobilization of As from soils and sediments. Indeed, recent experiments (Islam *et al.*, 2004) using As-rich sediments from West Bengal have demonstrated that stimulation of microbial Fe(III) reduction by the addition of acetate to slurry microcosms resulted in the release of As from the sediments. Initially, the released As was in the form of As(V), but this was subsequently reduced to As(III) showing that As(V) reduction was decoupled from Fe(III) reduction. The increased As release by the addition of a carbon source suggests that the production of toxic, mobile As(III) in sediments from the Ganges delta may be due, at least in part, to the influx of surface-derived carbon into deeper sediments and the resultant stimulation of anaerobic, metal-reducing bacteria.

Although most research involving microbial redox transformations of As has focused on As mobilization and As reduction, As(III) can also be microbially oxidized to As(V). This was first reported by Green in 1918 who described the isolation of an arsenite-oxidizer from an arsenical cattle-dipping solution (Green, 1918). There have since been a number of other reports of As(III)-oxidizers, most of which are heterotrophic or mixotrophic in that they require organic carbon for growth (Osborne and Ehrlich; 1976; Phillips and Taylor, 1976; vanden Hoven and Santini, 2004). Although the oxidation of As(III) by O_2 or nitrate/nitrite is exergonic, it has been suggested that these organisms perform the reaction for detoxification purposes and derive only maintenance energy from the reaction (Ehrlich, 1978). There has also been a recent report of a chemoautotrophic As(III)-oxidizer, capable of growth using O_2 as a terminal electron acceptor and bicarbonate as the carbon source (Santini et $al.$, 2000). The isolation of a facultative, chemoautotroph capable of coupling As(III)-oxidation with reduction of NO_2^- or NO_3^- suggests that bacteria are able to recycle microbially reduced As back to As(V) in sediments under both oxic and anoxic conditions (Oremland et $al.$, 2002).

Studies of bacterial Se redox transformations have, to some extent, focused on the dissimilatory reduction of the soluble oxyanions, SeO_4^{2-} and SeO_3^{2-}, to elemental, immobile Se^0 as a potential bioremediation technology for Se-contaminated soils and sediments. SeO_4^{2-} and SeO_3^{2-} can be reduced by many bacteria including representatives from the genera $Enterobacter$ (Losi and Frankenberger, 1997), $Pseudomonas$ (Lortie et $al.$, 1992), $Bacillus$ (Garbisu et $al.$, 1996), $Salmonella$ (McCready et $al.$, 1966), $Streptococcus$ (Tilton et $al.$, 1967), and $Desulfovibrio$ (Tomei et $al.$, 1995). Se^0 accumulates outside the cells rather than as internal precipitates. Although reduction of Se oxyanions in most cases is an anaerobic process, there are also reports of aerobic reduction. Aerobic reduction of SeO_4^{2-} and SeO_3^{2-} was reported for an isolate of $P.$ $stutzeri$ (Lortie et $al.$, 1992) and SeO_3^{2-} was reduced to Se^0 by aerobically grown $Salmonella$ $heidelberg$ (McCready et $al.$, 1966). Aerobic reduction may function as a detoxification mechanism or a cometabolic reaction with no direct benefit to the microorganism.

Selenite appears to be reduced by a larger number of bacteria than SeO_4^{2-}, but the latter compound has been shown to support anaerobic growth. Although selenium and sulfur are both Group VIB elements and one might expect microbial reduction of both to be catalyzed by similar bacteria and pathways, most of the Se-respiring organisms described in the literature do not utilize SO_4^{2-} as a terminal electron acceptor and the mechanisms of SeO_4^{2-} and SeO_3^{2-} reduction have not been fully elucidated for most organisms. In some cases, Se-reducers are also capable of nitrate respiration and complete Se reduction only occurs in the presence of nitrate (Macy, 1994; Losi and Frankenberger, 1997).

The long-term immobilization of selenium as Se^0 in soils is unlikely since Se^0 can be reoxidized both by abiotic reactions (Massacheleyn *et al.*, 1990) and microbial activity (Losi and Frankenberger, 1997). As described by Dungan and Frankenberger, uptake of SeO_4^{2-} and SeO_3^{2-} by plants, or microbial methylation and subsequent volatilization may be more effective in removal of high concentrations of Se from soil pools (Dungan and Frankenberger, 1999).

Microbial Transformations of Mercury

Most of the metals discussed in this chapter are enzymatically transformed to generate energy or to disperse excess reductant. The microbial transformations of Hg appear to primarily serve a detoxification function or to occur fortuitously, and would be beyond the scope of this review if not for the extreme toxicity of many Hg compounds and the critical role that microorganisms play in Hg speciation and cycling. Mercury is also noteworthy since Hg resistance is probably the most studied and best understood of the microbial metal resistances. A brief description of Hg cycling is therefore provided and the interested reader is referred to recent, in-depth reviews for additional information on microbial resistance and transformations of mercury (Baldi, 1997; Hobman *et al.*, 2000; Barkay *et al.*, 2003).

In addition to release from gold refining and various industrial processes and products, Hg is also produced in the gaseous form as a result of coal burning and waste incineration. In addition to its elemental form, Hg^0, environmental mercury exists in the mercurous (Hg_2^{2+} or Hg^+) or mercuric (Hg^{2+}) oxidation states or in the form of organic mercury compounds. The latter includes the highly toxic, (mono) methyl mercury (CH_3Hg^+), the somewhat less toxic dimethyl mercury (($CH_3)_2Hg$), and humate complexes (Jonasson, 1970). The speciation of Hg determines its mobility, bioavailability, biomagnification potential, and degree of toxicity. Hg^0 and ($CH_3)_2Hg$ are volatile and transformation to these relatively water-insoluble forms allows a potential reduction of local soil concentrations as the Hg moves into the atmospheric compartment. CH_3Hg^+ is soluble in water but also lipophilic, and is highly bioavailable and able to bioaccumulate and biomagnify in living tissues. The inorganic Hg^+ and Hg^{2+} cations may be mobile in the soil solution or may bind to negatively charged sites in soils.

The complex cycling of Hg between these different species is largely driven by microbial activity, although abiotic reactions can also play a significant role. It had previously been believed that CH_3Hg^+ was formed primarily by the fortuitous reaction between Hg^{2+} and the methylcobalamin present in methanogens and other anaerobic bacteria. Further methylation of CH_3Hg^+ by the same mechanism can produce $(CH_3)_2Hg$, although this second methylation step proceeds much more slowly (See Ehrlich, 1996c, for a review) More recent work, however, suggests that sulfate-reducing bacteria such as *Desulfovibrio desulfuricans* may be responsible for much of the methylation that occurs in anoxic estuarine and freshwater sediments and that the reaction is enzymatically driven (Compeau and Bartha, 1985; Gilmour *et al.*, 1992; Choi *et al.*, 1994). Other bacteria, including both aerobes and anaerobes, have also been shown to participate in mercury methylation (Robinson and Tuovinen, 1984).

In addition to methylation reactions, microorganisms also participate in other Hg transformations including methylmercury degradation. The well-studied mercury resistance (*mer*) operon in some bacteria encodes enzymes that confer a narrow-range resistance to inorganic mercury via reduction by a mercuric reductase to Hg^0 or, in other cases, broad-spectrum resistance to organomercury compounds via cleavage of the C-Hg bond by an organomercurial lyase and subsequent reduction of Hg(II) to Hg^0 (Begley *et al.*, 1986a; Begley *et al.*, 1986b). Alternately, a microbial oxidative demethylation pathway can lead to production of CO_2 and CH_4 and is considered the dominant demethylation route in some anoxic soils or sediments (Oremland *et al.*, 1995). Hydrogen sulfide produced by a strain of *D. desulfuricans* has also been shown to react with CH_3Hg^+ to form a dimethylmercury sulfide $[(CH_3Hg)_2S]$ which subsequently degrades abiotically to release $(CH_3)Hg$, metacinnabar (cubic HgS), methane, and small amounts of cationic Hg (Baldi *et al.*, 1993). The bacterial oxidation of Hg^0 to Hg(II) has also been recently demonstrated for some bacteria including strains of *Bacillus* and *Streptomyces* (Smith *et al.*, 1998).

INDIRECT, MICROBIALLY-MEDIATED ALTERATION OF METAL MOBILITY

Although one of foci of this review is the direct and indirect microbial redox transformation of metals and the effects on metal mobility in soils, it is important to emphasize that microbial activity can effect metal mobility through at least several other mechanisms. In heterogeneous soils, these mechanisms can clearly occur simultaneously and the *net* change in metal mobility can be a result of concomitant redox transformations, mineral dissolution or precipitation, pH alteration, metal complexation, mineral and microbial sorption, and other factors. The predominant mechanism of mobility change in a particular soil will largely be determined by the geochemical characteristics of the inorganic and organic phase, nature of the microbial community, pH, availability of oxygen and other bacterial electron acceptors, degree of water saturation, and many other factors.

In column studies using soil containing previously sorbed Cd^{2+}, for example, Chanmugathas and Bollag studied the effect of nutrient (glucose, sucrose, peptone, yeast extract, $NaNO_3$) amendment on Cd solubilization (Chanmugathas and Bollag, 1988). Nutrient addition increased 38-day Cd solubilization to 16% and 36%, respectively, in sterile and nonsterile columns from solubilization values of 6% and 9% in sterile and nonsterile columns without nutrient amendment. The increased Cd solubilization in nonsterile columns was attributed to microbial production of unidentified, low-molecular-weight organic compounds that were capable of forming soluble Cd-complexes. A reduction in pH and the formation of trace metal complexes with microbial metabolites was also proposed to explain the solubilization of Cd, Cu, Pb, and Zn oxides under anoxic conditions by a *Clostridium* species (Francis and Dodge, 1988). In the same studies, Francis and Dodge showed that Fe and Mn oxides underwent dissolution via direct enzymatic reduction by the *Clostridium* species, suggesting that different metals may be mobilized by different mechanisms even in a very simple system containing a single microbial species.

When Cd, Cr, Ni, Pb, and Zn were coprecipitated with the iron (hydr) oxide, goethite, rather than presented as a metal oxide, the same *Clostridium* species was able to remobilize the metals either during reductive dissolution of the oxide or via production of metabolites capable of forming soluble metal complexes (Francis and Dodge, 1990). Mn-, Co-, Cr-, and Al-substituted goethites were also shown to similarly release the co-precipitated trace metal during reductive dissolution of the iron (hydr)oxide by a species of *Clostridium butyricum* (Bousserrhine et al., 1999). The direct applicability of these last studies to circumneutral soil systems is arguable since the medium utilized contained a high concentration of fermentable substrate (5 to 10 g l^{-1} glucose plus lesser amounts of peptone or yeast extract) that wouldn't be encountered in soils and resulted in dramatic decreases in medium pH and production of organic acids. However, in subsequent studies done at circumneutral pH using the iron reducing bacterium, *S. putrefaciens* CN32, the release of coprecipitated Ni has been demonstrated during microbial reduction of hydrous Fe(III) (hydr)oxide (Fredrickson et al., 2000), as has the release of coprecipitated Co and Ni during reduction of goethite (Zachara et al., 2001).

Interestingly, other researchers have demonstrated that other trace metals sorbed to Fe (hydr)oxides can instead be immobilized within new mineral phases formed during microbial (hydr)oxide reduction (Cooper et al., 2000; Roden et al., 2002). When Fe(III) (hydr)oxides are microbially reduced in a soil, the Fe(II) produced can exist in various forms depending on local system geochemistry. In addition to being present as free Fe^{2+}, the ferrous ion may form soluble complexes with NOM or suitable inorganic ligands. Soluble species may sorb to the (hydr)oxide surface or other solid-phases, or may precipitate alone or as coprecipitates with other metals. In addition, discrete, new, Fe(II)-containing mineral phases may be formed, e.g., sorption of microbially-produced Fe(II) to lepidocrocite can result in formation of biogenic magnetite (Cooper et al., 2000). Depending on the iron phases present in the soil and the chemistry of the soil or sediment solution, biogenic minerals may include magnetite, siderite, green rusts, vivianite, and various other non-identified or metastable minerals (Fredrickson et al., 1998; Zachara et al., 1998; Cooper et al., 2000). Some of the biogenic, Fe(II)-containing phases are strong reductants that may have an indirect effect on the mobility of contaminant metals (discussed below).

The formation of these biogenic mineral phases may, even in the absence of further redox transformation of the contaminant metal, result in a decrease in metal mobility and toxicity by 'trapping' the metal in stable mineral matrices. It is believed that the free ion species is more bioavailable or toxic than ion pairs, polymers, soluble complexes, or sorbed species (Chaney, 1988). Microbial activities that cause long-term conversion or sequestration of trace metals into less-toxic species would therefore be expected to reduce the hazards associated with metals in soils. Although incorporation of metals into soluble or insoluble organic complexes may reduce toxicity, the decrease in toxicity may be relatively short-lived. Sludge-born trace metals, for example, may eventually be released as the organic portion of the complex is biodegraded. If the metals, however, are incorporated into the crystalline structure of a newly created, biogenic mineral, long-term reductions in toxicity, mobility, and bioavailability may result.

Cooper et al. (2000), for example, studied the fate of previously sorbed Zn (~ 300 μM) in slurries containing goethite or lepidocrocite and the dissimilatory iron-reducing bacterium, S. putrefaciens 200. As Fe(III) reduction proceeded in lepidocrocite-containing systems, there was a decrease in the amount of Zn that could be recovered by 0.5 M HCl extraction (pH ~ 0.3), even though the Zn could be recovered by digestion in 6 M HCl. X-ray diffraction analyses showed the conversion of lepidocrocite to magnetite and the immobilization of Zn was attributed to incorporation within the magnetite crystalline structure. Similar results, i.e., movement of Zn from a readily-extractable, sorbed phase to an immobilized phase insoluble in 0.5 M HCl, were observed in slurries containing goethite instead of lepidocrocite, although the identity of the biogenic mineral phase could not be clearly discerned in such systems. Such a process could have important consequences for the mobility of metals such as Zn in anoxic, iron-bearing soils and sediments. Indeed, it is possible that immobilization of some metals may be facilitated by the addition of suitable substrates to hasten Fe(III) reduction and trap contaminant metals in newly-formed, Fe(II)-

containing minerals. The long-term stability of such biogenic minerals remains to be determined, however, as does the impact of other soil components, e.g., clays and NOM, on the incorporation of metals into biogenic minerals. In similar studies using natural sediment slurries rather than synthetic goethite or lepidocrocite, for example, Cooper et al. (2005) observed limited immobilization of Zn during microbial reduction of Fe(III) in the sediment, suggesting that the identity and crystallinity of the iron-bearing phase and the presence of silicate clays as a competing sorbant may strongly affect the extent of contaminant metals immobilized during iron reduction and formation of new mineral phases. The demonstration that Sr could similarly be incorporated into biogenic Fe(II)-containing minerals during microbial reduction of an amorphous Fe(III) oxide confirmed that this immobilization mechanism is not limited to Zn and may be important for a number of trace metals in iron-bearing soils and sediments (Roden et al., 2002).

Overview of Indirect, Microbial Redox Transformations

In addition to the direct, enzymatic reduction of many metallic soil contaminants described above, bacteria are often able to indirectly transform soil pollutants via indirect reduction mechanisms. Production of Fe(II) by microbial reduction of Fe(III) (hydr)oxides, for example, can result in the subsequent abiotic reduction of a contaminant metal by the Fe(II). The Fe(III) formed as a product of such an abiotic reaction could potentially be again available for microbial reduction. In such a case, the iron would function as a redox mediator by shuttling electrons between the microorganism and the contaminant metal as depicted in Fig. 3.1. Such a mechanism could expand the number of metals that could be affected by microbial activity in a contaminated soil matrix, since it would not require that the bacterium be capable of enzymatically reducing the metal. Indeed, the establishment of zones of Fe(II)-containing minerals (permeable reactive barriers) by

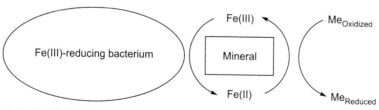

Fig. 3.1   Fe(III)/Fe(II) as a redox mediator for reduction of metals. The Fe(III) mineral is reduced by Fe(III) reducing bacteria and the Fe(II)-bearing product can then abiotically reduce the oxidized metal ($Me_{Oxidized}$). The Fe(III) formed during production of the reduced metal ($Me_{Reduced}$) is depicted as being subject to repeated microbial reduction.

stimulation of Fe(III)-reducing bacteria has been proposed to reduce horizontal migration of contaminant metals in the saturated zone (Fig. 3.2).

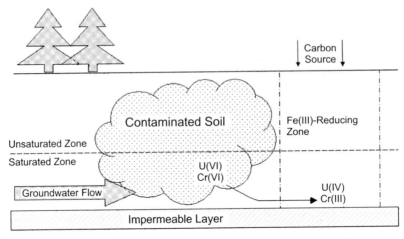

Fig. 3.2 Cross-sectional view of a permeable reactive biobarrier. The activity of Fe(III) reducing bacteria are stimulated by addition of a carbon source, e.g., lactate or acetate, creating a Fe(II)-containing, reducing zone. Soluble Cr(VI) or U(VI) transported with groundwater through this permeable reactive barrier are reduced to Cr(III) and U(IV) which are precipitated and removed from the groundwater flow.

Due to the ubiquity of iron (hydr)oxides, iron-bearing clays, and natural organic matter, these components are likely the most important redox mediators in most soils. The ability of NOM to serve as a redox shuttle or mediator is generally believed to be due to quinone moieties in humic matter, although bound transition metals and other redox-active functional groups in humic compounds also present opportunities for such reactions (Senesi and Miano, 1995). Although soil NOM and humic compounds are known to act as an electron shuttle for microbial iron reduction (Scott *et al.*, 1998; Lovley, 1999), the ability of soil NOM to increase rates of indirect microbial redox transformation of pollutants has been most extensively studied for organic pollutants. In studies with *S. putrefaciens* 200, for example, Backhus and coworkers showed that the rate and extent of carbon tetrachloride dechlorination was enhanced by a high-organic soil and suggested that compounds in the soil were catalyzing the transfer of electrons from the microorganism to the chlorinated compound (Backhus *et al.*, 1997), a mechanism later confirmed using humic compounds extracted from the soil (Collins and Picardal, 1999). The ability of soil NOM to function as an electron transfer mediator has also been demonstrated for other organic pollutants such as hexachloroethane (Curtis and Reinhard, 1994) and nitrobenzenes (Schwarzenbach *et al.*, 1990), and it is possible that microbially reduced humics can also serve as electron shuttles for a variety of reducible metals in soils.

Fe(II) has been clearly shown to be an important natural reductant that can potentially transform a number of organic and metallic pollutants in anaerobic soil and sediment systems. Microbially-produced Fe(II) in such systems may exist as aqueous Fe^{2+} in the soil solution, although it is more likely that aqueous Fe(II) would be in the form of aqueous complexes due to the presence of soluble NOM and other aqueous complexants. In most soil or sediment systems, however, one would expect most Fe(II) to be a component of the solid phase. The Fe(II) may be present as a structural component of microbially-reduced minerals, e.g., magnetite or ferruginous clays such as nontronite. In addition, the Fe(II) maybe complexed by NOM coatings on the mineral surfaces or sorbed to clays, iron oxides, or other mineral surfaces. The sorption may occur at ion-exchange sites such as the siloxane surface on clay minerals and consist of relatively weak, outer-sphere complexes. Alternately, sorption may occur at mineral sites, e.g., surface hydroxyl groups, that offer the opportunity for formation of stronger, inner-sphere complexes.

Microbially-produced Fe(II), in a variety of aqueous or solid-phase species, therefore presents a suite of reductants for the transformation of contaminant metals in soils and sediments. Aqueous Fe(II) species may reduce some metals in reactions in the soil solution (see Cr below), but reactions of metals with sorbed Fe(II) or with Fe(II)-containing minerals are likely more important in soils and sediments. Solid-phase Fe(II) is a stronger reducing agent than is aqueous Fe^{2+} and numerous studies have demonstrated that aqueous metals or organic contaminants can be reduced at elevated rates by sorbed or structural Fe(II). Sorbed Fe(II) or biogenic, Fe(II)-containing minerals have been identified as the reductant for a number of chlorinated organic pollutants in a number of studies suggesting that such reactions may be important transformation pathways in many contaminated soils (Kim and Picardal, 1999; Amonette *et al.*, 2000; Cervini-Silva *et al.*, 2001; McCormick *et al.*, 2002; O'Loughlin and Burris, 2004).

There is also significant evidence that microbially-produced Fe(II) in soils may function as an effective reductant for contaminant metals, especially when present as sorbed Fe(II) or Fe(II)-containing minerals. This has most clearly been shown for reactions of Fe(II) with Cr(VI) (described below), but other metallic contaminants in soils may be similarly transformed. Uranyl $(U(VI)O_2^{2+})$ was reduced by Fe(II) when sorbed to the surface of hematite or natural lake sediments whereas no reduction was observed by aqueous Fe^{2+} (Liger *et al.*, 1999). Structural Fe(II) in magnetite $[Fe(II)Fe(III)_2O_4]$ and ilmenite $[Fe(II)TiO_3]$, for example, was able to reduce aqueous chromate, cupric, ferric, and vanadate ions over a wide pH range (White and Peterson, 1996). As depicted in Figure 3.1, therefore, microbially-reduced iron in soils may serve as a renewable reservoir of reductant for reductive transformations of metals not known to be directly reduced by microorganisms.

Indirect Transformations of Chromium

The possibility of chemical Cr(VI) reduction by a variety of inorganic and organic species in soils requires that both biotic and abiotic processes be simultaneously considered when studying the fate and transport of Cr in soils and sediments. In addition to the direct microbial reduction of highly toxic and soluble Cr(VI) to Cr(III) as described above, iron-reducing bacteria are able to promote Cr(VI) reduction by production of Fe(II). The biogenic Fe(II) is subsequently able to abiotically reduce Cr(VI), often at rates that exceed direct, enzymatic reduction (Wielinga $et\ al.$, 2001).

$$3\ Fe^{2+} + HCrO_4^- + 8\ H_2O\ 3\ Fe(OH)_3 + Cr(OH)_3 + 5\ H^+$$

The ability of aqueous Fe^{2+} to reduce Cr(VI) is well known and increases with pH for a given Fe(II) concentration (Sedlak and Chan, 1997; Schlautman and Han, 2001). Carboxylates and phenolates present in soil NOM may accelerate reaction rates but can also potentially mobilize Cr(III) by leading to the formation of soluble Cr(III) complexes (Buerge and Hug, 1998). Interestingly, although the reduction rate of Cr(VI) by Fe(II) decreases at low pH, studies using undisturbed soil columns revealed that the abiotic reduction rates of Cr(VI) by NOM bound to soil mineral surfaces was quite rapid at low pH (pH ~ 4) (Jardine $et\ al.$, 1999). In those studies, Cr(III) was immobilized on soil particulates and Cr(VI) had a relatively short half-life of 85 hr.

The ability of mineral-sorbed Fe^{2+} to chemically reduce Cr(VI) has been demonstrated for several iron (hydr)oxides, clays, and natural, loamy sands (Buerge, 1999). Cr(VI) reduction has also been observed in studies with Fe(II)-bearing minerals such as biotite (Ilton and Veblen, 1994), Fe(II)-containing goethite (Bidoglio $et\ al.$, 1993), and the Fe(II)-Fe(III) green rusts present in some hydromorphic soils (Loyaux-Lawniczak $et\ al.$, 2000). The chemical reaction between Fe^{2+} or Fe(II)-bearing minerals was also used to explain the reductive immobilization of Cr(VI) in soils of a contaminated industrial waste landfill (Loyaux-Lawniczak $et\ al.$, 2001).

The Fe(II) produced during microbial reduction of Fe(III)-bearing minerals can also be the driving force for similar reactions in soils. As Fe(II) is produced by DIRBs, much of the Fe(II) will sorb to mineral surfaces and provide a reactive reductant. Wielinga and coworkers showed that the DIRB, $Shewanella\ alga$ strain BrY was able to promote the reduction of Cr(VI) in such a fashion (Wielinga $et\ al.$, 2001). In slurries of $S.\ alga$ and the Fe(III) (hydr)oxides, goethite, hematite, ferrihydrite, and HFO, Cr(VI) was reduced much more rapidly than by bacterial suspensions alone. In addition, their work suggested that the $Fe(OH)_3$ produced during Cr(VI) reduction by biogenic Fe(II) could be reduced again by *S. alga*, effectively acting as an electron shuttle between the bacterium and the chromate. Since such an indirect reduction mechanism may minimize Cr toxicity that results from direct microbial Cr(VI) reduction (described above), more Cr(VI) might be reduced indirectly via production of Fe(II) than via direct enzymatic reduction. Additional work is necessary, but it is clear that coupled biotic-abiotic reactions may be very important for immobilization of Cr in iron-reducing soils and sediments.

Immobilization of Other Metals During Reoxidation of Fe(II)

In soils containing elevated concentrations of heavy metal pollutants, the sorptive capacity of soil solids may be exceeded and the metals may be mobile in the soil solution. If aqueous Fe^{2+} is also present, microbial Fe^{2+} oxidation by FeOB will generate Fe(III) (hydr)oxides that may immobilize the metals by co-precipitation or adsorption. Lack et al, for example, recently demonstrated that anaerobic nitrate-dependent oxidation of Fe(II) by *Dechloromonas* species resulted in the precipitation of Fe(III) (hydr)oxides and rapid adsorption and removal of 55 μM U(VI) and 81 μM Co(II) from solution (Lack *et al.*, 2002). Abiotic oxidation of biogenic Fe^{2+} by atmospheric oxygen may similarly be able to immobilize aqueous metals in soils that undergo redox fluctuations due to alternating periods of saturation and drying. In experiments designed to mimic such anoxic-oxic conditions, slurries of Zn-contaminated goethite and natural sediments were oxidized by exposure to air following an anoxic period of microbial iron reduction. Precipitation of iron (hydr)oxides resulted in a decrease in the concentrations of aqueous Zn^{2+} (Cooper and Picardal, 2000). The long-term stability of such metal-containing Fe(III) (hydr)oxides is currently unknown and much work is necessary to determine if immobilization of metals during stimulated reoxidation of Fe(III) can be used as an effective soil bioremediation strategy.

MICROBIAL EFFECTS ON ASSOCIATIONS OF METALS WITH SOIL MINERALS

Much of the recent research into microbially induced changes in metal speciation has focused on the ability of microorganisms to directly or indirectly alter metal oxidation state. However, it is important to emphasize that these transformations are only one of many ways in which bacteria can alter metal speciation. In addition to adsorbing metals onto cell surfaces via surface complexation and biomineralization processes, soil microorganisms can also induce geochemical changes that affect both metal adsorption and metal co-precipitation with common soil minerals. This section will discuss the general biogeochemical processes by which soil microorganisms alter metal association with soil minerals other than iron oxides. Metal adsorption and precipitation at microbial surfaces, as well as the effect of microbial exudates on metal speciation are discussed in subsequent sections.

Aluminosilicates

Aluminosilicates, or clay minerals, comprise an important component of all soil systems and their geochemistry has been extensively documented. With respect to metal biogeochemistry, these minerals provide important surfaces that act as an attachment site for microorganisms and adsorb metallic contaminants. Metal adsorption to clay mineral surfaces can occur via cation exchange, i.e., nonspecific formation of outer-sphere surface complexes, or through site-specific, chemical reactions that produce inner-sphere surface complexes. Outer-sphere interactions prevail for Group I/II cations (e.g. Na^+, K^+, Ca^{2+}, Mg^{2+}, Sr^{2+}), but the mechanism of interaction is more nebulous for transition metals that can adsorb via either mechanism. This complexity arises from the chemical and physical properties of the diffuse layer of cations surrounding a clay mineral surface. These properties change non-linearly with pH, ionic strength, redox state of transition metals in the crystalline matrix, presence/absence of organic matter, saturation state of carbonate minerals, and numerous other system parameters. Microbial activity can simultaneously alter many (if not all) of these properties, and investigations into these potentially coupled effects are challenging and valuable. This section offers a brief review of what is known about the effects of microbial iron reduction on aluminosilicate geochemistry, and how these changes may be expected to alter metal chemistry in soil systems.

Microbial reduction of structural Fe(III) in clay minerals can dramatically alter clay mineral structure and reactivity. Wu and coworkers reported that microbial reduction of structural Fe(III) in sodium nontronite gels reduced swelling pressure at intermediate oxidation states (Wu et al., 1988). It has also been demonstrated that reduction of structural Fe(III) in standard smectite clays by indigenous *Pseudomonas* strains decreased both swelling pressures and water content (Gates et al., 1993). These processes are reversible (Gates et al., 1996), decrease particle surface area while increasing surface charge density (Kostka et al., 1999b), and may produce a larger particle-size and pore-size distribution (Gates et al., 1998). Dong and coworkers also recently demonstrated that microbial Fe(III) reduction can dissolve smectite clays (Dong et al., 2003a), and alter the morphology of illite clays (Dong et al., 2003b). It has also been reported that structural Fe(III) in low grade kaolin can be reduced by enrichment cultures of dissimilatory iron reducing bacteria (Lee et al., 2002). Addition of organic ligands increases the extent of Fe(III) reduction (Kostka *et al.*, 1999a), and enhance these effects. Bacteria incapable of dissimilatory FeIII reduction can also alter clay mineral chemistry. *Pseudomonas mendocina*, and assimilatory Fe(III)-reducing bacteria incapable of dissimilatory Fe(III) reduction, for example, can obtain nutrient iron from kaolinite clays that contain less than 1% iron by mass (Maurice *et al.*, 2001a, b), indicating that a broad range of soil bacteria may be able to modify clay mineral reactivity.

Relatively little is known about the effect of these microbial alterations of clay mineral reactivity on metal geochemistry in soils, but current knowledge does suggest four mechanisms by which microbial activity can alter clay mineral reactivity towards metals: clay mineral dissolution, production of organic exudates, alteration of surface charge, and alteration of surface potential. Zachara and coworkers have demonstrated that many contaminant metals adsorb strongly to edge sites on 2:1 clays (e.g. smectite) (Zachara and McKinley, 1993; Zachara and Smith, 1994; Zachara *et al.*, 2002), and it has been shown that microbial Fe(III) reduction can dissolve these edge sites (Dong *et al.*, 2003a). Thus, this process can reduce the affinity of contaminant metals for clay mineral surfaces in soils. A number of soil microorganisms produce organic exudates such as siderophores and exopolysaccharides (EPS), and organic matter has been reported to alter the rate, extent, and mechanism of metal uptake by clay minerals (Takahashi *et al.*, 1999; Neubauer *et al.*, 2000; Nachtegaal and Sparks, 2003). These studies generally indicate that organic matter either enhances the rate and extent of metal uptake or has minimal effect at circumneutral pH. In all cases, however, organic matter makes adsorbed metals more susceptible to acid facilitated desorption. These changes in soil metal chemistry are thought to result from a balance between organic-induced changes in surface charge and the formation of metal-organic complexes. Kraepiel *et al.* (1999) incorporated these processes into a mechanistic model for metal adsorption onto montmorillonite, whereby surface charge and surface potential affect metal adsorption by both cation exchange and surface complexation mechanisms. Microbial reduction of smectite Fe(III) reduces surface area while increasing surface charge (Kostka *et al.*, 1999b), thereby altering the balance of chemical processes that control metal sorption onto smectite clays. By extension, microbial reduction of Fe(III) in illite and kaolinite clays may have a similar effect. Finally, microbial Fe(III) reduction can reduce clay mineral surface potential through generation of structural and surficial Fe(II) (Ernstsen *et al.*, 1998; Hofstetter *et al.*, 2003).

Very few studies have directly examined the effect of microbial processes on metal geochemistry in clays and clayey soils. Marsh *et al.* (2000) reported that microorganisms in clay-like aquifer sediments can reduce Cr(VI), but did not specifically investigate the effects of clay minerals. Neubauer and coworkers reported that the siderophore desferrioxamine B (DFOB) reduced surface charge and dramatically increased metal sorption to Na-montmorillonite and kaolinite at circumneutral pH (Neubauer *et al.*, 2000). Conversely, DFOB increased metal mobility at elevated pH. Cooper *et al.* (2004) reported that microbial reduction of Fe(III) in a iron-bearing clayey sediment resulted in the transfer of previously adsorbed Zn from Zn-O surface complexes on sedimentary iron oxide minerals to outer-sphere $ZnCl_2$ complexes likely associated with clay minerals. These initial studies demonstrate that microbe-aluminosilicate interactions can dramatically alter metal geochemistry in soil environments, and indicate that further investigations into this issue would provide valuable new knowledge.

Carbonates

The carbonate system forms the geochemical backbone of many environmental processes. Solid and aqueous carbonate species place a primary control on solution chemistry, and carbonate concretions (or cements) are an important physical and chemical component of soils and sediments. Carbonates can alter metal partitioning and speciation in many ways. Carbonate minerals control solution pH in many natural systems, and some metals form strong aqueous complexes with carbonate anions. More importantly, perhaps, many metals partition into the crystal structure of carbonate minerals to varying degrees (Veizer, 1990). The degree of metal co-precipitation with carbonate minerals is partially dependent on both the metal's chemistry (e.g. ionic radius, free energy of the metal ion in solution) and the rate of mineral precipitation and mineral saturation state (Davis *et al.*, 1987; Morse, 1990; Veizer, 1990; Tesoriero and Pankow, 1996; Curti, 1999). Carbonate mineral precipitation and dissolution can also affect metal adsorption onto sedimentary oxide, silicate, and aluminosilicate minerals by other mechanisms. Davranche and coworkers have demonstrated that the presence of carbonates can exert a strong influence on the surface charge of natural minerals, and therefore on metal surface complexation chemistry (Davranche *et al.*, 2002). Significant changes in the pH_{znpc} (zero proton surface charge) for natural soils can be seen with as little as 1 wt% $CaCO_3$. Carbonate mineral nucleation can also act to cement sediment grains together, allowing dissolution and precipitation of these minerals to alter reactive surface area of sediments (Verrecchia, 1994; Wang *et al.*, 1994).

Many researchers have investigated the processes by which microorganisms can incorporate metals into carbonate minerals. This literature is extensive, and generally examines direct biologically mediated carbonate mineral precipitation rather than indirect, microbiologically induced processes (Veizer, 1990; Fortin *et al.*, 1997; Dove *et al.*, 2003). With respect to indirect microbiolgically induced processes, Ehrlich has identified five conditions that can result in carbonate mineral production: (1) oxidation of organic compounds in a buffered neutral or alkaline environment; (2) oxidation of organic nitrogen compounds to produce ammonia and CO_2 in unbuffered environments; (3) dissimilatory SO_4^{2-} reduction; (4) CO_2 removal from bicarbonate solutions; and (5) urea hydrolysis (Ehrlich, 1996d). All of these processes can affect the degree to which a metal is incorporated into calcite minerals and the ability of carbonate minerals to affect the surface reactivity of other sedimentary phases. With respect to soil processes, most researchers in this field have investigated the ability to utilize microbial urea hydrolysis to induce calcite precipitation in the subsurface. Urea hydrolyzing bacteria are widespread in the environment (Mobley and Hausinger, 1989; Fujita *et al.*, 2000), and several researchers have investigated various innovative applications for this process (Stocks-Fischer *et al.*, 1999; Warren *et al.*, 2001; Hammes *et al.*, 2003; Fujita *et al.*, 2004). Of these, only Warren *et al.* (2001) and Fujita *et al.* (2004) have investigated the application of this process to the problem of metal remediation in groundwater and soils. These researchers have demonstrated that microbiologically induced calcite precipitation can rapidly remove strontium from solution, and that this process is also capable of removing copper and uranium. Copper uptake was weak, due to Cu toxicity to the microorganism used. Fujita *et al.* (2004) have also demonstrated that artificially enhanced urea hydrolysis by *B. pasteurii* can enhance calcite precipitation and incorporate Sr^{2+} into the calcite crystal matrix. Faster calcite precipitation rates yielded higher Sr/Ca ratios. These initial results indicate that calcite precipitation is a common environmental process facilitated by microorganisms, and establish a strong foundation for future research in this area. It should also be noted that microbial production of CO_2 and organic acids in soil environments can shift conditions to favor carbonate dissolution, and reverse the effects of carbonate precipitation on metal geochemistry.

Phosphates

Phosphate minerals are not widely found in soil environments in high concentrations, but phosphate addition is a common technique for *in situ* metal remediation. The most common approach is to add apatite to the system, either as a permeable reactive barrier or as a soil amendment. For example, Cao *et al.* (2003) reported that PO_4^{3-} amendment of soil at a metal contaminated site effectively immobilized Pb within a pyromorphite-like mineral while slightly reducing Cu and Zn mobility. Similarly, Chen *et al.* (1997) reported that Pb incorporated within PO_4^{3-} minerals was much more resistant to standard toxicity-test leaching procedures than Cd and Zn that was only adsorbed to the surface of added hydroxyapatite ($Ca_5(PO_4)_3OH$). Field studies have also indicated that PO_4^{3-} minerals can significantly alter metal mobility. It has been reported that naturally occurring PO_4^{3-} minerals can greatly retard uranium transport (Jerden *et al.*, 2003), and that phosphoric acid addition can be used to remediate Pb contamination at a smelter site (Yang *et al.*, 2001). The success of H_3PO_4 addition in stimulating *in situ* Pb remediation suggests that other PO_4^{3-} generating processes may also act to immobilize contaminant metals. Macaskie and coworkers demonstrated that the enzymatic degradation of organic phosphate compounds by *Citrobacter* spp can be linked to Am, Np, Pu, and U phosphate mineral precipitation in laboratory systems (Macaskie *et al.*, 1996).

The ability to utilize externally supplied phosphate ester compounds in order to acquire phosphate for cellular growth and metabolism is widespread in the prokaryotic world, and up to 70 to 80% of a soil microbial population may be able to hydrolyze organic phosphate esters with phosphatases (Kertesz *et al.*, 1994). Phosphatase activity can occur under both aerobic and anaerobic conditions (Landis, 1991; Pehkonen and Zhang, 2002), and addition of phosphate esters has been shown to enhance degradation rates of organic contaminants in field demonstrations (Palumbo *et al.*, 1995). There is ample evidence to suggest that microbial P-cycling may impact metal geochemistry in soil environments, but very few studies have investigated this possibility. In a lonely foray into this field, Jensen and coworkers have demonstrated that abiotic phytic acid decomposition generates PO_4^{3-} minerals that have the capability to sequester actinide and lanthanide metals (Jensen *et al.*, 1996). We are not aware of any other published investigations into the potential for microbial degradation of organic phosphorus compounds to reduce metal mobility. It is important to note that PO_4^{3-} minerals also provide a source of nutrient-P in the subsurface and are susceptible to microbially enhanced dissolution. Welch *et al.* (2002) demonstrated that soil microorganisms can facilitate PO_4^{3-} release from apatite minerals without altering solution pH, and Taunton *et al.* (2000) reported that soil bacteria and fungal hyphae act to solubilize phosphate in weathered soils. Thus, while soil microorganisms can potentially be stimulated to immobilize metals within phosphate minerals, they may also facilitate slow leaching of metals through release of organic acids and "mining" of nutrient-phosphorus.

Sulfides

Metal sulfides are an important class of minerals that have a profound effect on biogeochemical cycles and can be generated by either microbial or physical processes (Vaughan and Lennie, 1991). Research into the interactions between metals, microorganisms, and sulfide minerals is extensive, and has historically revolved around two topics; acid mine drainage, and metal-sulfur cycling in aquatic sediments. Investigations into the effect of microbial S-cycling on metal geochemistry in soil and groundwater environments have been largely limited to studies investigating processes that affect SO_4^{2-} reduction rates and the ability for cells to act as nucleation sites for sulfide minerals.

The unique ability for sulfide minerals to immobilize many contaminant metals is driven both by the strong adsorption affinity of iron sulfides for many trace metals (Jean and Bancroft, 1986; Morse and Arakaki, 1993), and the extremely low solubility product of many metal sulfide minerals (Morse and Arakaki, 1993; Cooper and Morse, 1998). Metals remain immobilized within sulfide minerals while reducing conditions persist, but can be remobilized during periodic oxidation events (Morse, 1994; Cooper and Morse, 1998; Morse and Rickard, 2004). This mineralization and immobilization process is driven by microbial production of hydrogen sulfide, which generally requires high levels of organic carbon. Once the rate of soil NOM oxidation exceeds the rate of O_2 transport, soil microorganisms utilize alternative electron acceptors, including SO_4^{2-}. Recent studies have demonstrated that predominantly anaerobic processes such as sulfate reduction can occur in bulk aerobic systems (Parkin, 1987; Postma and Boesen, 1991) and that the occurrence of anaerobic microsites is largely responsible for this phenomenon (Arah, 1990; Smith, 1990; Sakita and Kusuda, 2000). Most soil environments do not contain sufficient sulfate or organic carbon to stimulate extensive SO_4^{2-} reduction, but there are a few notable exceptions. Cozzarelli and coworkers have reported the occurrence of SO_4^{2-} reduction in aquifers contaminated by gasoline and/or landfill leachate (Cozzarelli et al., 1999; Cozzarelli et al., 2000). Kennedy and Everett (2001) have demonstrated that gypsum addition to landfill microcosms promoted SO_4^{2-} reduction over methanogenesis and promoted sulfide mineral formation while reducing CH_4 emissions. It has been observed that microbial SO_4^{2-} reduction in a landfill clay liner can attenuate aqueous Zn via the formation of zinc sulfides (Kamon et al., 2002), and several researchers have demonstrated that a compost and wood/mulch based permeable reactive barriers can effectively remove heavy metals from heavily contaminated groundwater (Benner et al., 2002; Ludwig et al., 2002). Thus, sulfide mineral formation can be an important process governing metal geochemistry in contaminated groundwater and soil environments, where diffusive fluxes of O_2 are limited and organic carbon and sulfate are sufficient.

BIOSORPTION AND BIOMINERALIZATION ON MICROBIAL SURFACES

Bacterial surfaces contain a high density of organic amide, carboxyl, and phosphoryl groups. These surface functional groups are highly reactive towards metals, and can alter metal geochemistry by three general mechanisms: surface complexation with microbial surfaces, co-precipitation with minerals that nucleate on microbial surfaces, and surface complexation with minerals that nucleate on microbial surfaces. In a soil environment, it is extremely difficult to discriminate between metal surface complexation with microbial surfaces and metal surface complexation with minerals associated with a cellular surface. This section will not attempt discuss the separate effects of these two surface complexation processes on metal geochemistry in soils, and will instead focus on the different implications of surface complexation and co-precipitation of metals at or near a bacterial surfaces.

It is important to note that the ratio of microbial surface area to mineral surface area may generally be too low to allow a significant effect from direct metal sorption to bacterial surfaces. For example, at an estimated cellular surface area of 140 m^2 g^{-1} cells (Fein *et al.*, 1997), a soil cell density that ranges from 10^5 to 10^8 cells g^{-1} soil would produce approximately 4×10^{-6} to 4×10^{-3} m^2 of cellular surface area per gram of soil. Most soils have a BET surface area that ranges from 5 to 50 m^2 g^{-1}, and the microbial surface area in most soils would therefore be about 1000 to 1,000,000 times smaller than the mineral surface area. Given this low ratio, it is natural to wonder why the topic of metal sorption to microbial surfaces has attracted a great deal of recent interest in environmental research. Besides the obvious allure of the unknown, the answer may lie in the fact that there are several unique properties of bacteria that allow them to have an impact far greater than their small size would predict. First, microbial surfaces are not static. Bacteria are continuously generating new cellular and exo-polymeric surfaces and their time-integrated impact can be much greater than that of a single "snapshot". Second, metal interactions with microbial surfaces can alter microbial metabolism. Third, metal adsorption to bacterial surfaces often leads to mineral nucleation and metal incorporation into the crystal structure of a new soil mineral. Fourth, various combinations of these processes can lead to important alterations in fluid and bacterial transport through soil environments. The following sections briefly summarize some important aspects of these processes by which microbial surfaces alter metal geochemistry in soil environments.

Metal Adsorption to Microbial Surfaces (Biosorption)

Bacterial surfaces are known to strongly adsorb a number of metals, and these phenomena have been the topic of over 50 years of research. Some of the earliest studies examined the interactions of uranium with cellular surfaces (Dounce and Flagg, 1949; Rothstein and Meier, 1951), and were eventually followed by examinations of the potential to use bacteria as industrial adsorbents and studies of the role of metal-binding surfaces in environmental systems (Brady and Duncan, 1994; Kovacova and Sturdik,

2002). Most of the environmental work has highlighted the primary importance of carboxyl and phosphoryl groups in metal binding (Beveridge and Fyfe, 1985; Yee and Fein, 2001; Yee and Fein, 2003), though hydroxyl groups can be important at elevated pH (Fein *et al.*, 1997; Daughney *et al.*, 1998) and amine groups can be important for anion sorption (Urrutia-Mera and Beveridge, 1993). The surface site density, protonation constants, and metal stability constants can vary slightly between different species (Daughney *et al.*, 1998), a phenomenon that may be related to differences in charge distribution throughout the entire cell wall of a microorganism (Sokolov *et al.*, 2001). However, the effects of these complex charge structures are generally not large, and some researchers have suggested that a universal model may describe non-metabolic metal sorption to all bacteria (Yee and Fein, 2001; Yee and Fein, 2003).

These studies indicate that bacterial surfaces can adsorb appreciable quantities of metals, and that current knowledge allows us to model this process. Yet, it is still unclear how important this process is to metal cycling in soil environments. Ledin and coworkers have demonstrated that microorganisms that contribute < 2% of total soil mass can adsorb over 35% of total added Zn, Cd, and Hg (Ledin *et al.*, 1999). Similarly, it has been demonstrated that aerobic microbial activity in soils reduces Cd availability, even though it was not clear if this was an adsorptive process or metabolic effect (Kurek and Bollag, 2004). In contrast, several researchers have noted that microorganisms can also facilitate metal transport through sandy soils (Guine *et al.*, 2003; Jensen-Spaulding *et al.*, 2004), through both exopolysaccharide generation and direct cellular adsorption and transport. Yee and Fein (2002) have demonstrated that, depending upon the degree of bacterial transport, Cd adsorption to bacterial cells can either facilitate or inhibit Cd transport through soil columns. These studies indicate that metal adsorption onto cellular surfaces can significantly alter metal transport in soil environments, and that the magnitude and direction of this change is a factor of microbial metabolism and microbial affinity for a given soil environment. Full quantification of these microbe-metal vectors is difficult, as metal contaminants are known to alter microbial metabolism (Suarez and Reyes, 2002) and the metabolic processes that control bacterial attachment to mineral surfaces are not well understood.

Mineral Precipitation on Microbial Surfaces (Biomineralization)

Mineral nucleation and formation on bacterial surfaces is a well known phenomenon that has been extensively documented (Fortin *et al.*, 1997; Douglas and Beveridge, 1998). Direct microbially-mediated and indirect, microbially-induced processes have been described for both iron biomineralization (Bazylinski and Moskowitz, 1997; Konhauser, 1997; 1998) and carbonate biomineralization (Fortin *et al.*, 1997; Dove *et al.*, 2003).

Indirect, microbially-induced biomineralization has been described for a broad range of minerals, including aluminosilicates, phosphates, silicates, and sulfides (Beveridge *et al.*, 1983; Macaskie *et al.*, 1996; Fortin *et al.*, 1997; Konhauser and Urrutia, 1999; Perry, 2003). The nature of these microbiological biominerals is diverse, as is their ability to impact metal cycling within soil systems. The geochemistry of these biominerals is similar to that of their abiotic analogs, with one exception. Microbial biominerals are universally small, with a correspondingly high surface area/mass ratio and a proclivity to co-precipitate trace metals within the crystal matrix. Cell metabolism can also dramatically alter the region immediately surrounding the cell, generating geochemical conditions that are quite different from bulk solution (Fortin *et al.*, 1997; Little *et al.*, 1997). These unique conditions can generate kinetic effects and geochemical anisotropy that affect metal cycling in soil environments.

While the ability of bacterial surfaces to stimulate mineral nucleation is well known, the complex and dynamic geochemistry of the cellular environment makes this process very difficult to model, especially in heterogeneous soil environments. Thus, most studies into the effect of biomineralization on metal cycling have been phenomenonalistic in nature. In addition to the previous discussions of carbonate and sulfide mineral biomineralization, a number of strict and facultative anaerobic bacteria are known to metabolically reduce a number of metals (e.g. U(VI), Cr(VI), and Tc(VI)) to produce insoluble oxides that nucleate on cellular surfaces (Lovley and Coates, 1997). Microbial Fe(III) reduction has been reported to sequester Sr within siderite ($FeCO_3$) (Parmar *et al.*, 2000), and Zn within goethite (Cooper *et al.*, 2000; Cooper *et al.*, 2004). In addition, it has been shown that microbial cell surfaces enhance $Cu(OH)_2$ precipitation over cell-free controls (Fowle and Fein, 2001). These biomineralization processes generally act to sequester metals, but are exceedingly difficult to separate from biosorption. Future research needs to better discriminate between these processes, and link the relative extent of metal sequestration to various measures of soil microbial activity.

CONCLUSIONS

The widespread contamination of soils by heavy metals requires a logical approach to predicting metal fate and transport, assessing the risks associated with the contamination, and developing remediation methods to mitigate unacceptable risks to human health or the environment. The high costs associated with excavation, transport, and disposal of contaminated soils in a secure landfill make it likely that a large proportion of contaminated soils will have to remain in place and be treated *in situ*. Regardless of whether such treatment is physical, chemical, or biological in nature; the contaminant metals will remain in place and be subjected to a wide range of microbiological and geochemical processes that that affect metal mobility, bioavailability, and toxicity. Successful management of this long-term risk will require detailed knowledge of the complex biogeochemical factors that control metal fate and transport, and a variety of techniques that can be used to control metal transport as necessary. This task requires description and quantification of microbiological processes that affect metal geochemistry in soil environments. A mechanistic description of these processes in soils, however, is complicated by physical, chemical, and biological heterogeneities across all size scales, from the nanometer to the kilometer.

Soil microorganisms can impact metal geochemistry by a wide range of direct and indirect mechanisms. Direct effects include biosorption, production of metal chelating agents, and direct metabolic alteration of metal oxidation state. These direct pathways are relatively straightforward, and many researchers believe that some of these processes can eventually be harnessed to immobilize metals *in situ*. Indirect effects arise from microbiological interactions with the surrounding soil environment, and are much more difficult to quantify. Indirect processes include alteration of metal oxidation state by reactions with biogenic Fe(II) or other microbially generated reductant, alteration of aluminosilicate surface chemistry via microbial Fe(III) reduction, precipitation of carbonates by bacterial ureolysis, dissolution of carbonates via microbial CO_2 production, formation of sulfide minerals via microbial SO_4^{2-} reduction, microbial mediation of the soil P-cycle, and mineral nucleation on bacterial surfaces. While more complex, these indirect processes must be considered and quantified when managing the long-term fate of a metal contaminated site. Direct remediation of a metal-contaminated soil can be expensive, and cannot be maintained indefinitely. Once active remediation is complete, the indirect processes may be likely to control metal fate and transport as the system slowly returns to a less manipulated and more natural state.

REFERENCES

Amonette, J.E., Workman, D.J., Kennedy, D.W., Fruchter, J.S., Gorby, Y.A. (2000). Dechlorination of carbon tetrachloride by Fe(II) associated with goethite. Environmental Science and Technology 34: 4606-4613.

Arah, J.R.M. (1990). Diffusion-reaction models of denitrification in soil microsites. In :Denitrification in Soil and Sediment. pp. 245-258. Eds. N.P. Revsbech, N.P. and J. Sørensen. Plenum Press, New York.

Arnold, R.G., DiChristina, T.J., Hoffmann, M.R. (1988). Reductive dissolution of Fe(III) oxides by *Pseudomonas* sp. 200. Biotechnology and Bioengineering 32: 1081-1096.

Atkinson, M., Bayer, P., Drell, D., Faison, B., Hazen, T., Katz, A., McCullough, J., Palmisano, A., Seybold, S., Wuy, L. (2003). Bioremediation of metals and radionuclides... What it is and how it works. A NABIR primer. Office of Science, U.S. Department of Energy, Berkeley, CA, pp. 79.

Backhus, D., Picardal, F.W., Johnson, S., Knowles, T., Collins, R., Kim, S. (1997). Soil- and surfactant-enhanced reductive dechlorination of carbon tetrachloride in the presence of *Shewanella putrefaciens* 200. Journal of Contaminant Hydrology 28: 337-361.

Badar, U., Ahmed, N., Beswick, A.J., Pattanapipitpaisal, P., Macaskie, L.E. (2000). Reduction of chromate by microorganisms isolated from metal contaminated sites of Karachi, Pakistan. Biotechnology Letters 22: 829-836.

Baldi, F. 1997. Microbial transformation of mercury species and their importance in the biogeochemical cycle of mercury. In: Metal Ions in Biological Systems. Vol. 34, pp. 213-257. Ed. A. Sigel. Marcel Dekker, New York.

Baldi, F., Pepi, M., Filippelli, M. (1993). Methylmercury resistance in *Desulfovibrio desulfuricans* strains in relation to methylmercury degradation. Applied and Environmental Microbiology 59: 2479-2485.

Barkay, T., Miller, S.M., Summers, A.O. (2003). Bacterial mercury resistance from atoms to ecosystems. FEMS Microbiology Reviews 27: 355-384.

Bazylinski, D.A., Moskowitz, B.M. (1997). Microbial biomineralization of magnetic iron minerals: microbiology, magnetism and environmental significance. In: *Geomicrobiology*, Vol. 35, pp. 181-224. Eds. J.F. Banfied, and K.H. Nealson. Mineralogical Society of America, Washington D. C.

Begley, T.P., Walts, A.E., Walsh, C.T. (1986a). Bacterial organomercurial lyase: Overproduction, isolation, and characterization. Biochemistry 25: 7186-7192.

Begley, T.P., Walts, A.E., Walsh, C.T. (1986b). Mechanistic studies of a protonolytic organomercurial cleaving enzyme: Bacterial organomercurial lyase. Biochemistry 25: 7192-7200.

Beliaev, A.S., Saffarini, D.A., McLaughlin, J.L., Hunnicutt, D. (2001). MtrC, an outer membrane decahaem *c* cytochrome required for metal reduction in *Shewanella putrefaciens* MR-1. Molecular Microbiology 39: 722-730.

Benner, S.G., Blowes, D.W., Ptacek, C.J., Mayer, K.U. (2002). Rates of sulfate reduction and metal sulfide precipitation in a permeable reactive barrier. Applied Geochemistry 17: 301-320.

Benz, M., Brune, A., Schink, B. (1998). Anaerobic and aerobic oxidation of ferrous iron at neutral pH by chemoheterotrophic nitrate-reducing bacteria. Archives of Microbiology 169: 159-165.

Beveridge, T.J., Fyfe, W.S. (1985). Metal fixation by bacterial cell walls. Canadian Journal of Earth Science 22: 1893-1898.

Beveridge, T.J., Meloche, J.D., Fyfe, W.S., Murray, R.G.E. (1983). Diagenesis of metals chemically complexed to bacteria: laboratory formation of metal phosphates, sulfides, and organic condensates in artificial sediments. Applied and Environmental Microbiology 45: 1094-1108.

Bidoglio, G., Gibson, P.N., O'Gorman, M., Roberts, K.J. (1993). X-ray absorption spectroscopy investigation of surface redox transformations of thallium and chromium on colloidal mineral oxides. Geochimica et Cosmochimica Acta 57: 2389-2394.

Blencowe, D.K., Morby, A.P. (2003). Zn(II) metabolism in prokaryotes. FEMS Microbiology Reviews 27: 291-311.

Blum, J.S., Bindi, A.B., Buzzelli, J., Stolz, J.F., Oremland, R.S. (1998). *Bacillus arsenicoselenatis* sp. nov., and *Bacillus seleitireducens*, sp. nov.: two haloalkaliphiles from

Mono Lake, California that respire oxyanions of selenium and arsenic. Archives of Microbiology 171: 19-30.

Bopp, H.L., Ehrlich, H.L. (1988). Chromate resistance and reduction in *Pseudomonas fluorescens* strain LB300. Archives of Microbiology 150: 426-431.

Bousserrhine, N., Gasser, U.G., Jeanroy, E., Berthelin, J. (1999). Bacterial and chemical reductive dissolution of Mn-, Co-, Cr-, and Al-substituted goethites. Geomicrobiology Journal 16: 245-258.

Brady, D., Duncan, J.R. (1994). Bioaccumulation of metal-cations by *Saccharomyces cerevisiae*. Applied Microbiology and Biotechnology 41: 149-154.

Brock, T.D., Gustafson, J. (1976). Ferric iron reduction by sulfur-and iron-oxidizing bacteria. Applied and Environmental Microbiology 32: 567-571.

Buerge, I.J. (1999). Influence of mineral surfaces on chromium (VI) reduction by iron (II). Environmental Science and Technology 33: 4285-4291.

Buerge, I.J., Hug, S.J. (1998). Influence of organic ligands on chromium(VI) reduction by iron(II). Environmental Science and Technology 32: 2092-2099.

Cao, R.X., Ma, L.Q., Chen, M., Singh, S.P., Harris, W.G. (2003). Phosphate-induced metal immobilization in a contaminated site. Environmental Pollution 122: 19-28.

Cervini-Silva, J., Larson, R.A., Wu, J., Stucki, J.W. (2001). Transformation of chlorinated aliphatic compounds by ferruginous smectite. Environmental Science and Technology 35: 805-809.

Chaney, R.L. (1988). In: Metal Speciation: Theory, Analysis, and Application. Eds. J.R. Kramer, and H.E. Allen, Lewis Publishers, Inc., Chelsea.

Chanmugathas, P., Bollag, J.-M. (1988). A column study of the biological mobilization and speciation of cadmium in soil. Archives of Environmental Contamination and Toxicology 17: 229-237.

Chen, X.B., Wright, J.V., Conca, J.L., Peurrung, L.M. (1997). Evaluation of heavy metal remediation using mineral apatite. Water Air and Soil Pollution 98: 57-78.

Choi, S., Chase, T., Bartha, R. (1994). Enzymatic catalysis of mercury methylation by Desulfovibrio desulfuricans LS. Applied and Environmental Microbiology 60: 1342-1346.

Collins, R., Picardal, F.W. (1999). Enhanced anaerobic transformations of carbon tetrachloride by soil organic matter. Environmental Toxicology and Chemistry 18: 2703-2710.

Compeau, G., Bartha, R. (1985). Sulfate-reducing bacteria: principal methylators of mercury in anoxic estuarine sediments. Applied and Environmental Microbiology 50: 1203-1207.

Cooper, D.C., Morse, J.W. (1998). Biogeochemical controls on trace metal cycling in anoxic marine sediments. Environmental Science and Technology 32: 327-330.

Cooper, D.C., Neal, A.L., Kukkadapu, R.K., Brewe, D., Coby, A.J., Picardal, F.W. 2005. Effects of sediment iron mineral composition on microbially mediated changes in divalent metal speciation: importance of ferrihydrite. Geochimica et Cosmochimica Acta (In Press).

Cooper, D.C., Picardal, F., Rivera, J., Talbot, C. (2000). Zinc immobilization and magnetite formation via ferric oxide reduction by *Shewanella putrefaciens* 200. Environmental Science and Technology 34 (1): 100-106.

Cooper, D.C., Picardal, F.W. (2000). Influence of sequential microbial iron reduction and chemical oxidation on zinc speciation in contaminated natural sediments. Presented at the Annual Meeting of the Society of Environmental Toxicology and Chemistry Nashville, TN, USA.

Cornell, R.M., Schwertmann, U. (1996). The Iron Oxides, VCH, Weinheim.

Cozzarelli, I.M., Herman, J.S., Baedecker, M.J., Fischer, J.M. (1999). Geochemical heterogeneity of a gasoline-contaminated aquifer. Journal of Contaminant Hydrology 40: 261-284.

Cozzarelli, I.M., Suflita, J.M., Ulrich, G.A., Harris, S.H., Scholl, M.A., Schlottmann, J.L., Christenson, S. (2000). Geochemical and microbiological methods for evaluating anaerobic processes in an aquifer contaminated by landfill leachate. Environmental Science and Technology 34: 4025-4033.

Cummings, D.E. (1999). Arsenic mobilization by the dissimilatory Fe(III)-reducing bacterium *Shewanella alga* BrY. Environmental Science and Technology 33 (5): 723-729.

Curti, E. (1999). Coprecipitation of radionuclides with calcite: estimation of partition coefficients based on a review of laboratory investigations and geochemical data. Applied Geochemistry 14: 433-445.

Curtis, G.P., Reinhard, M. (1994). Reductive dehalogenation of hexachloroethane, carbon tetrachloride, and bromoform by anthrahydroquinone disulfonate and humic acid. Environmental Science and Technology 28: 2393-2401.

Daughney, C.J., Fein, J.B., Yee, N. (1998). A comparison of the thermodynamics of metal adsorption onto two common bacteria. Chemical Geology 144: 161-176.

Davis, J.A., Fuller, C.C., Cook, A.D. (1987). A model for trace-metal sorption processes at the calcite surface: Adsorption of Cd^{2+} and subsequent solid-solution formation. Geochimica et Cosmochimica Acta 51: 1477-1490.

Davison, W., Seed, G. (1983). The kinetics of the oxidation of ferrous iron in synthetic and natural waters. Geochimica et Cosmochima Acta 47: 67-79.

Davranche, M., Beaufreton, S., Bollinger, J.C. (2002). Influence of carbonates on the surface charge of a natural solid. Journal of Colloid and Interface Science 249: 113-118.

Dong, H.L., Kostka, J.E., Kim, J. (2003a). Microscopic evidence for microbial dissolution of smectite. Clays and Clay Minerals 51: 502-512.

Dong, H.L., Kukkadapu, R.K., Fredrickson, J.K., Zachara, J.M., Kennedy, D.W., Kostandarithes, H.M. (2003b). Microbial reduction of structural Fe(III) in illite and goethite. Environmental Science and Technology 37: 1268-1276.

Douglas, S., Beveridge, T.J. (1998). Mineral formation by bacteria in natural microbial communities. FEMS Microbiology Ecology 26: 79-88.

Dounce, A.L., Flagg, J.F. (1949). The chemistry of uranium compounds. In : Pharmacology and Toxicology of Uranium Compounds, part I, pp. 55-145. Eds. C. Voegtlin, H.C. Hodge. McGraw-Hill, New York.

Dove, P.M., De Yoreo, J.J., Weiner, S. (2003). Biomineralization. Mineralogical Society of America, Washington D.C.

Dungan, R.S., Frankenberger, W.T., Jr. (1999). Microbial transformations of selenium and the bioremediation of seleniferous environments. Bioremediation Journal 3: 171-188.

Ehrenreich, A., Widdel, F. (1994). Anaerobic oxidation of ferrous iron by purple bacteria, a new type of phototrophic metabolism. Applied and Environmental Microbiology 60: 4517-4526.

Ehrlich, H.L. (1978). Inorganic energy sources for chemolithotrophic and mixotrophic bacteria. Geomicrobiology Journal 1: 65-83.

Ehrlich, H.L. (1996a). Geomicrobiology of iron. In : Geomicrobiology, pp. 312-388. Ed, H.L. Ehrlich. Marcel Dekker, Inc., New York.

Ehrlich, H.L. (1996b). Geomicrobiology of managanese. In : Geomicrobiology, pp. 389-489. Ed, H.L. Ehrlich. Marcel Dekker, Inc., New York.

Ehrlich, H.L. (1996c). Geomicrobiology of mercury. In : Geomicrobiology, pp. 295-311. Ed, H.L. Ehrlich. Marcel Dekker, Inc., New York.

Ehrlich, H.L. (1996d). Microbial formation and degradation of carbonates. In: Geomicrobiology, pp. 172-216. Ed, H.L. Ehrlich. Marcel Dekker, Inc., New York.

Emerson, D. (2000). Microbial oxidation of Fe(II) and Mn(II) at circumneutral pH. In: Environmental Microbe-Metal Interactions, pp. 31-52. Ed, D.R. Lovley. ASM Press, Washington, D.C.

Emerson, D., Moyer, C. (1997). Isolation and characterization of novel iron-oxidizing bacteria that grow at circumneutral pH. Applied and Environmental Microbiology 63: 4784-4792.

Ernstsen, V., Gates, W.P., Stucki, J.W. (1998). Microbial reduction of structural iron in clays - A renewable source of reduction capacity. Journal of Environmental Quality 27: 761-766.

Fein, J.B., Daughney, C.J., Yee, N., Davis, T.A. (1997). A chemical equilibrium model for metal adsorption onto bacterial surfaces. Geochimica et Cosmochimica Acta 61: 3319-3328.

Fortin, D., Ferris, F.G., Beveridge, T.J. (1997). Surface-mediated mineral development by bacteria. In: Geomicrobiology, Vol. 35, pp. 161-181. Eds, J.F. Banfied and K.H. Nealson. Mineralogical Society of America, Washington D.C.

Fowle, D.A., Fein, J.B. (2001). Quantifying the effects of *Bacillus subtilis* cell walls on the precipitation of copper hydroxide from aqueous solution. Geomicrobiology Journal 18: 77-91.

Francis, A.J., Dodge, C.J. (1988). Anaerobic microbial dissolution of transition and heavy metal oxides. Applied and Environmental Microbiology 54: 1009-1014.

Francis, A.J., Dodge, C.J. (1990). Anaerobic microbial remobilization of toxic metals coprecipitated with iron. Environmental Science and Technology 24: 373-378.

Fredrickson, J.K., Zachara, J.M., Kennedy, D.W., Dong, H., Onstott, T.C., Hinman, N.W., Li, S.-M. (1998). Biogenic iron mineralization accompanying the dissimilatory reduction of hydrous ferric oxide by a groundwater bacterium. Geochimica et Cosmochimica Acta 62: 3239-3257.

Fredrickson, J.K., Zachara, J.M., Kukkadapu, R.K., Gorby, Y.A., Smith, S.C., Brown, C.F. (2000). Biotransformation of Ni-substituted hydrous ferric oxide by an Fe(III)-reducing bacterium. Environmental Science and Technology 35: 703-712.

Fude, L., Harris, B., Urrutia, M., Beveridge, T. (1994). Reduction of Cr(VI) by a consortium of sulfate-reducing bacteria (SRB III). Applied and Environmental Microbiology 60: 1525-1531.

Fujita, Y., Ferris, E.G., Lawson, R.D., Colwell, F.S., Smith, R.W. (2000). Calcium carbonate precipitation by ureolytic subsurface bacteria. Geomicrobiology Journal 17: 305-318.

Fujita, Y., Redden, G.D., Ingram, J.C., Cortez, M.M., Ferris, F.G., Smith, R.W. (2004). Strontium incorporation into calcite generated by bacterial ureolysis. Geochimica et Cosmochimica Acta 68: 3261-3270.

Gadd, G.M., Sayer, J.A. (2000). Influence of fungi on the environmental mobility of metals and metalloids. In: Environmental Microbe-Metal Interactions, pp. 237-256. Ed, D.R. Lovley. ASM Press, Washington, D.C.

Garbisu, C., Alkorta, I., Llama, M.J., Serra, J.L. (1998). Aerobic chromate reduction by *Bacillus subtilis*. Biodegradation 9: 133-141.

Garbisu, C., Ishii, T., Leighton, T., Buchanan, B.B. (1996). Bacterial reduction of selenite to elemental selenium. Chemical Geology 132: 199-204.

Gates, W.P., Jaunet, A.M., Tessier, D., Cole, M.A., Wilkinson, H.T., Stucki, J.W. (1998). Swelling and texture of iron-bearing smectites reduced by bacteria. Clays and Clay Minerals 46: 487-497.

Gates, W.P., Stucki, J.W., Kirkpatrick, R.J. (1996). Structural properties of reduced Upton montmorillonite. Physics and Chemistry of Minerals 23: 535-541.

Gates, W.P., Wilkinson, H.T., Stucki, J.W. (1993). Swelling properties of microbially reduced ferruginous smectite. Clays and Clay Minerals 41: 360-364.

Gilmour, C.C., Henry, E.A., Mitchell, R. (1992). Sulfate stimulation of mercury methylation in freshwater sediments. Environmental Science and Technology 26: 2281-2285.

Green, H.H. (1918). Description of a bacterium which oxidizes arsenite to arsenate, and of one which reduces arsenate to arsenite, isolated from a cattle-dipping tank. South African Journal of Science 14: 465-467.

Guine, V., Martins, J., Gaudet, J.P. (2003). Facilitated transport of heavy metals by bacterial colloids in sand columns. Journal de Physique IV 107: 593-596.

Hallbeck, L., Pederson, K. (1991). Autotrophic and mixotrophic growth of *Gallionella ferruginea*. Journal of General Microbiology 137: 2657-2661.

Hallbeck, L., Ståhl, F., Pedersen, K. (1993). Phylogeny and phenotypic characterization of the stalk-forming and iron-oxidizing bacterium *Gallionella ferruginea*. Journal of General Microbiology 139: 1531-1535.

Hammes, F., Boon, N., Clement, G., de Villiers, J., Siciliano, S.D., Verstraete, W. (2003). Molecular, biochemical and ecological characterisation of a bio-catalytic calcification reactor. Applied Microbiology and Biotechnology 62: 191-201.

Hanert, H.H. (2004a). The genus *Gallionella*. In: The Prokaryotes: An Evolving Electronic Resource for the Microbiological Community, 3rd edition, Release 3.0, 5/21/99; Ed, M. Dworkin. Springer-Verlag, New York. http://link.springer-ny.com/link/service/books/10125/.

Hanert, H.H. (2004b). The genus *Siderocapsa* (and other iron- or manganese-oxidizing Eubacteria. In: The Prokaryotes: An Evolving Electronic Resource for the Microbiological Community, 3rd edition, Release 3.0, 5/21/99; Ed, M. Dworkin, M. Springer-Verlag, New York, http://link.springer-ny.com/link/service/books/10125/.

Harrison Jr., A.P. (1984). The acidophilic thiobacilli and other acidophilic bacteria that share their habitat. Annual Review of Microbiology 38: 265-292.

Heijman, C.G., Holliger, C., Glaus, M.A., Schwarzenbach, R.P., Zeyer, J. (1993). Abiotic reduction of 4-chloronitrobenzene to 4-chloroaniline in a dissimilatory iron-reducing enrichment culture. Appl Environ Microbiol 59: 4350-4353.

Hobman, J.L., Wilson, J.R., Brown, N.L. (2000). Microbial mercury reduction. In: Environmental Microbe-Metal Interactions, pp. 177-197. Ed, D.R. Lovley. ASM Press, Washington, D.C.

Hoeft, S.E., Kulp, T.R., Stolz, J.F., Hollibaugh, J.T., Oremland, R.S. (2004). Dissimilatory arsenate reduction with sulfide as electron donor: Experiments with Mono Lake water and isolation of strain MLMS-1, a chemoautotrophic arsenate respirer. Applied and Environmental Microbiology 70: 2741-2747.

Hofstetter, T.B., Schwarzenbach, R.P., Haderlein, S.B. (2003). Reactivity of Fe(II) species associated with clay minerals. Environmental Science and Technology 37: 519-528.

Ilton, E.S., Veblen, D.R. (1994). Chromium sorption by phlogopite and biotite in acidic solutions at 25[deg]C: Insights from X-ray photoelectron spectroscopy and electron microscopy. Geochimica et Cosmochimica Acta 58: 2777-2788.

Islam, F.S., Gault, A.G., Boothman, C., Polya, D.A., Charnock, J.M., Chatterjee, D., Lloyd, J.R. (2004). Role of metal-reducing bacteria in arsenic release from Bengal delta sediments. Nature 430: 68-71.

Jardine, P.M., Fendorf, S.E., Mayes, M.A., Larsen, I.L., Brooks, S.C., Bailey, W.B. (1999). Fate and transport of hexavalent chromium in undisturbed heterogeneous soils. Environmental Science and Technology 33: 2939-2944.

Jean, G.E., Bancroft, G.M. (1986). Heavy-metal adsorption by sulfide mineral surfaces. Geochimica et Cosmochimica Acta 50: 1455-1463.

Jensen, M.P., Nash, K.L., Morss, L.R., Appelman, E.H., Schmidt, M.A. (1996). Immobilization of actinides in geomedia by phosphate precipitation.In: Humic and Fulvic Acids, Vol. 651, pp. 272-285. American Chemical Society, Washington.

Jensen-Spaulding, A., Shuler, M.L., Lion, L.W. (2004). Mobilization of adsorbed copper and lead from naturally aged soil by bacterial extracellular polymers. Water Research 38: 1121-1128.

Jentschke, G., Godbold, D.L. (2000). Metal toxicity and ectomycorrhizas. Physiologia Plantarum 109: 107-116.

Jerden, J.L., Sinha, A.K., Zelazny, L. (2003). Natural immobilization of uranium by phosphate mineralization in an oxidizing saprolite-soil profile: Chemical weathering of the Coles Hill uranium deposit, Virginia. Chemical Geology 199: 129-157.

Jonasson, I.P. (1970). Mercury in the natural environment: a review of recent work. Paper 70-57. Geological Survey of Canada, pp. 39.

Kamon, M., Zhang, H.Y., Katsumi, T. (2002). Redox effects on heavy metal attenuation in landfill clay liner. Soils and Foundations 42: 115-126.

Kappler, A., Newman, D.K. (2004). Formation of Fe(III)-minerals by Fe(II)-oxidizing photoautotrophic bacteria. Geochimica et Cosmochimica Acta 68: 1217-1226.

Kaufmann, F., Lovley, D.R. (2001). Isolation and characterization of a soluble NADPH-dependent Fe(III) reductase from Geobacter sulfurreducens. Journal of Bacteriology 183: 4468-4476.

Kennedy, L.G., Everett, J.W. (2001). Microbial degradation of simulated landfill leachate: Solid iron/sulfur interactions. Advances in Environmental Research 5: 103-116.

Kertesz, M.A., Cook, A.M., Leisinger, T. (1994). Microbial metabolism of sulfur-containing and phosphorus-containing xenobiotics. FEMS Microbiology Reviews 15: 195-215.

Kim, S., Picardal, F.W. (1999). Enhanced anaerobic biodegradation of carbon tetrachloride in the presence of reduced iron oxides. Environmental Toxicology and Chemistry 18: 2142-2150.

Konhauser, K.O. (1997). Bacterial iron biomineralisation in nature. FEMS Microbiology Reviews 20: 315-326.

Konhauser, K.O. (1998). Diversity of bacterial iron mineralization. Earth Science Reviews 43: 91-121.

Konhauser, K.O., Urrutia, M.M. (1999). Bacterial clay authigenesis: a common biogeochemical process. Chemical Geology 161: 399-413.

Kostka, J.E., Haefele, E., Viehweger, R., Stucki, J.W. (1999a). Respiration and dissolution of iron(III)-containing clay minerals by bacteria. Environmental Science and Technology 33: 3127-3133.

Kostka, J.E., Wu, J., Nealson, K.H., Stucki, J.W. (1999b). The impact of structural Fe(III) reduction by bacteria on the surface chemistry of smectite clay minerals. Geochimica et Cosmochimica Acta 63: 3705-3713.

Kovacova, S., Sturdik, E. (2002). Interactions between microorganisms and heavy metals including radionuclides. Biologia 57: 651-663.

Kraepiel, A.M.L., Keller, K., Morel, F.M.M. (1999). A model for metal adsorption on montmorillonite. Journal of Colloid and Interface Science 210: 43-54.

Kucera, S., Wolfe, R.S. (1957). A selective enrichment method for *Gallionella ferruginea*. Journal of Bacteriology 74: 344-349.

Kurek, E., Bollag, J.-M. (2004). Microbial immobilization of cadmium resleased from CdO in the soil. Biogeochemistry 69: 227-239.

Lack, J.G., Chaudhuri, S.K., Kelly, S.D., Kemner, K.M., O'Connor, S.M., Coates, J.D. (2002). Immobilization of radionuclides and heavy metals through anaerobic bio-oxidation of Fe(II). Applied and Environmental Microbiology 68: 2704-2710.

Landis, W.G. (1991). Distribution and nature of the aquatic organophosphorus acid anhydrases - enzymes for organophosphate detoxification. Reviews in Aquatic Sciences 5: 267-285.

Ledin, M., Krantz-Rulcker, C., Allard, B. (1999). Microorganisms as metal sorbents: comparison with other soil constituents in multi-compartment systems. Soil Biology and Biochemistry 31: 1639-1648.

Lee, E.Y., Cho, K.S., Ryu, H.W. (2002). Microbial refinement of kaolin by iron-reducing bacteria. Applied Clay Science 22: 47-53.

Liger, E., Charlet, L., Van Cappellen, P. (1999). Surface catalysis of uranium(VI) reduction by iron(II). Geochimica et Cosmochima Acta 63: 2939-2955.

Little, B.J., Wagner, P.A., Lewandowski, Z. (1997). Spatial relationships between bacteria and mineral surfaces. In Geomicrobiology: Interactions between Microbes and Minerals, Vol. 35, pp. 123-155. Eds, J. Banfield and K. Nealson. Mineralogical Society of America, Washington D.C.

Liu, C., Gorby, Y.A., Zachara, J.M., Fredrickson, J.K., Brown, C.F. (2002). Reduction kinetics of Fe(III), Co(III), U(VI), Cr(VI), and Tc(VII) in cultures of dissimilatory metal-reducing bacteria. Biotechnology and Bioengineering 80: 638-649.

Lloyd, J.R., Sole, V.A., Praagh, C.V.G.V., Lovley, D.R. (2000a). Direct and Fe(II)-mediated reduction of technetium by Fe(III)-reducing bacteria. Appl Environ Mircobiol 66: 3743-3749.

Lloyd, J.R., Yong, P., Macaskie, L.E. (2000b). Biological reduction and removal of Np(V) by two microorganisms. Environmental Science and Technology 34: 1297-1301.

Lortie, L., Gould, W.D., Rajan, S., McCready, R.G.L., Cheng, K.-J. (1992). Reduction of selenate and selenite to elemental selenium by a *Pseudomonas stutzeri* isolate. Applied and Environmental Microbiology 63: 3079-3084.

Losi, M.E., Frankenberger, W.T., Jr. (1997). Reduction of selenium oxyanions by *Enterobacter cloacae* SLDa-1: Isolation and growth of the bacterium and its expulsion of selenium particles. Applied and Environmental Microbiology 63: 3079-3084.

Lovley, D.R. (1993). Dissimilatory metal reduction. Annual Review of Microbiology 47: 263-290.

Lovley, D.R. (1999). Role of humic-bound iron as an electron transfer agent in dissimilatory Fe(III) reduction. Applied and Environmental Microbiology 65 : 4252-4254.

Lovley, D.R. (2000). Fe(III) and Mn(IV) reducton. In : Environmental Microbe-Metal Interactions, pp. 3-30. Ed, D.R. Lovley. ASM Press, Washington, D.C.

Lovley, D.R., Coates, J.D. (1997). Bioremediation of metal contamination. Current Opinion in Biotechnology 8: 285-289.

Lovley, D.R., Coates, J.D., Blunt-Harris, E.L., Phillips, E.J.P., Woodward, J.C. (1996). Humic substances as electron acceptors for microbial respiration. Nature 382: 445-448.

Lovley, D.R., Fraga, J.L. (1998). Humic substances as a mediator for microbially catalyzed metal reduction. Acta Hydrochim Hydrobiology 26: 152-157.

Lovley, D.R., Phillips, E.J.P. (1994). Reduction of chromate by Desulfovibrio vulgaris and its c_3 cytochrome. Applied and Environmental Microbiology 60: 726-728.

Lovley, D.R., Phillips, E.J.P., Gorby, Y.A., Landa, E.R. (1991). Microbial reduction of uranium. Nature 350: 413-416.

Lovley, D.R., Stolz, J.F., Nord, G.L., Jr. (1987). Anaerobic production of magnetite by a dissimilatory iron-reducing microorganism. Nature 330: 252-254.

Loyaux-Lawniczak, S., Lecomte, P., Ehrhardt, J.-J. (2001). Behavior of hexavalent chromium in a polluted groundwater: Redox processes and immobilization in soils. Environmental Science and Technology 35: 1350-1357.

Loyaux-Lawniczak, S., Refait, P., Ehrhardt, J.-J., Lecomte, P., Genin, J.-M.R. (2000). Trapping of Cr by formation of ferrihydrite during the reduction of chromate ions by Fe(II)—Fe(III) hydroxysalt green rusts. Environmental Science and Technology 34: 438-443.

Ludwig, R.D., McGregor, R.G., Blowes, D.W., Benner, S.G., Mountjoy, K. (2002). A permeable reactive barrier for treatment of heavy metals. Groundwater 40: 59-66.

Macaskie, L.E., Lloyd, J.R., Thomas, R.A.P., Tolley, M.R. (1996). The use of microorganisms for the remediation of solutions contaminated with actinide elements, other radionuclides, and organic contaminants generated by nuclear fuel cycle activities. Nuclear Energy-Journal of the British Nuclear Energy Society 35: 257-271.

Macy, J.M. (1994). Biochemistry of selenium metabolism by Thauera selenatis gen. nov. sp. nov. and use of the organism for bioremediation of selenium oxyanions in San Joaquin Valley drainage water. In : Selenium in the Environment, pp. 421-444. Eds, W.T. Jr. Frankenberger and S. Benson. Marcel Dekker, In., New York.

Magnuson, T.S., Hodges-Myerson, A.L., Lovley, D.R. (2000). Characterization of a membrane-bound NADH-dependent Fe^{3+}-reducing bacterium Geobacter sulfurreducens. FEMS Microbiology Letters 185: 205-211.

Marsh, T.L., Leon, N.M., McInerney, M.J. (2000). Physiochemical factors affecting chromate reduction by aquifer materials. Geomicrobiology Journal 17: 291-303.

Massacheleyn, P.H., Delaune, R.D., Patrick Jr, W.H. (1990). Transformations of selenium as affected by sediment oxidation-reduction potential and pH. Environmental Science and Technology 24: 91-98.

Maurice, P.A., Vierkorn, M.A., Hersman, L.E., Fulghum, J.E. (2001a). Dissolution of well and poorly ordered kaolinites by an aerobic bacterium. Chemical Geology 180: 81-97.

Maurice, P.A., Vierkorn, M.A., Hersman, L.E., Fulghum, J.E., Ferryman, A. (2001b). Enhancement of kaolinite dissolution by an aerobic Pseudomonas mendocina bacterium. Geomicrobiology Journal 18: 21-35.

McCormick, M.L., Bouwer, E.J., Adriaens, P. (2002). Carbon tetrachloride transformation in a model iron-reducing culture: Relative kinetics of biotic and abiotic reactions. Environmental Science and Technology 36: 403-410.

McCready, R.G.L., Campbell, J.N., Payne, J.I. (1966). Selenite reduction by Salmonella heidelberg. Canadian Journal of Microbiology 12: 703-714.

McLean, J., Beveridge, T.J. (2001). Chromate reduction by a pseudomonad isolated from a site contaminated with chromated copper arsenate. Applied and Environmental Microbiology 67: 1076-1084.

Mergeay, M., Monchy, S., Vallaeys, T., Auquier, V., Benotmane, A., Bertin, P., Taghavi, S., Dunn, J., van der Lelie, D., Wattiez, R. (2003). Ralstonia metallidurans, a bacterium specifically adapted to toxic metals: towards a catalogue of metal-responsive genes. FEMS Microbiology Reviews 27: 385-410.

Mobley, H.L.T., Hausinger, R.P. (1989). Microbial ureases - significance, regulation, and molecular characterization. Microbiology Reviews 53: 85-108.

Morse, J.W. (1990) The kinetics of calcium carbonate dissolution and precipitation. In Carbonates: Mineralogy and Chemistry, Vol. 11, pp. 399. Ed, R.J. Reeder. Mineralogical Society of America, Washington D.C.

Morse, J.W. (1994). Release of toxic metals via oxidation of authigenic pyrite in resuspended sediments. In : The Environmental Geochemistry of Sulfide Oxidation, pp. 289-297. Eds. C.N. Alpers and D.W. Blowes. American Chemical Society, Washington D.C.

Morse, J.W., Arakaki, T. (1993). Adsorption and co-precipitation of divalent metals with mackinawite (FeS). Geochimica et Cosmochimica Acta 57: 3635-3640.

Morse, J.W., Rickard, D. (2004). Chemical dynamics of sedimentary acid volatile sulfide. Environmental Science and Technology 38: 131A-136A.

Mulrooney, S.B., Hausinger, R.P. (2003). Nickel uptake and utilization by microorganisms. FEMS Microbiology Reviews 27: 239-261.

Myers, C.R., Myers, J.M. (1992). Localization of cytochromes to the outer membrane of anaerobically grown shewanella putrefaciens MR-1. Journal of Bacteriology 174: 3429-3438.

Myers, C.R., Myers, J.M. (1993). Role of menaquinone in the reduction of fumarate, nitrate, iron(III) and manganese(IV) by Shewanella putrefaciens MR-1. FEMS Microbiology Letters 114: 215-222.

Myers, C.R., Myers, J.M. (1997). Outer membrane cytochromes of Shewanella putrefaciens MR-1: spectral analysis, and purification of the 83-kDa c-type cytochrome. Biochimica et Biophysica Acta 1326: 307-318.

Nachtegaal, M., Sparks, D.L. (2003). Nickel sequestration in a kaolinite-humic acid complex. Environmental Science and Technology 37: 529-534.

Nealson, K.H., Myers, C.R. (1992). Microbial reducation of manganese and iron: New approaches to carbon cycling. Applied and Environmental Microbiology 58: 439-443.

Nealson, K.H., Saffarini, D. (1994). Iron and manganese in anaerobic respiration: Environmental significance, physiology, and regulation. Annual Review of Microbiology 48: 311-43.

Neubauer, U., Nowack, B., Furrer, G., Schulin, R. (2000). Heavy metal sorption on clay minerals affected by the siderophore desferrioxamine B. Environmental Science and Technology 34: 2749-2755.

Nevin, K.P., Lovley, D.R. (2000). Lack of production of electron-shuttling compounds or solubilization of Fe(III) during reduction of insoluble Fe(III) oxide by Geobacter metallireducens. Applied and Environmental Microbiology 66: 2248-2251.

Nevin, K.P., Lovley, D.R. (2002). Mechanisms for accessing insoluble Fe(III) oxide during dissimilatory Fe(III) reduction by Geothrix fermentans. Applied and Environmental Microbiology 68: 2294-2299.

Newman, D., Beveridge, T., Morel, F. (1997). Precipitation of arsenic trisulfide by Desulfotomaculum auripigmentum. Applied and Environmental Microbiology 63: 2022-2028.

Newman, D.K., Kolter, R. (2000). A role for excreted quinones in extracellular electron transfer. Nature 405: 94-97.

Ohlendorf, H.M., Hoffman, D.J., Saiki, M.K., Aldrich, T.W. (1986). Embryonic mortality and abnormalities of aquatic birds: Apparent impact of selenium from irrigation drain water. Science of the Total Environment 52: 49-63.

Ohtake, H., Fujii, E., Toda, T. (1990). A survey of effective electron donors for reduction of toxic hexavalent chromate by *Enterobacter* cloacae (strain HO1). Journal of General and Applied Microbiology 36: 203-208.

O'Loughlin, E.J., Burris, D.R. (2004). Reduction of halogenated ethanes by green rust. Environmental Toxicology and Chemistry 23: 41-48.

Oremland, R., Miller, L., Dowdle, P., Connell, T., Barkay, T. (1995). Methylmercury oxidative degradation potentials in contaminated and pristine sediments of the Carson River, Nevada. Applied and Environmental Microbiology 61: 2745-2753.

Oremland, R.S., Hoeft, S.E., Santini, J.M., Bano, N., Hollibaugh, R.A., Hollibaugh, J.T. (2002). Anaerobic oxidation of arsenite in Mono Lake water and by a facultative, arsenite-oxidizing chemoautotroph, strain MLHE-1. Applied and Environmental Microbiology 68: 4795-4802.

Oremland, R.S., Stolz, J. (2000). Dissimilatory reduction of selenate and arsenate in nature. In : Environmental Microbe-Metal Interactions, , pp. 199-224, Ed, D.R. Lovley. ASM Press, Washington, D.C.

Osborne, F.H., Ehrlich, H.L. (1976). Oxidation of arsenite by a soil isolate of Alcaligenes. Journal of Applied Bacteriology 41: 295-305.

Palumbo, A.V., Scarborough, S.P., Pfiffner, S.M., Phelps, T.J. (1995). Influence of nitrogen and phosphorus on the in-situ bioremediation of trichloroethylene. Applied Biochemistry and Biotechnology 51-2: 635-647.

Parkin, T.B. (1987). Soil microsites as a source of denitrification variability. Soil Science Society of America Journal. 51: 1194.

Parmar, N., Warren, L.A., Roden, E.E., Ferris, F.G. (2000). Solid phase capture of strontium by the iron reducing bacteria *Shewanella alga* strain BrY. Chemical Geology 169: 281-288.

Pehkonen, S.O., Zhang, Q. (2002). The degradation of organophosphorus pesticides in natural waters: A critical review. Critical Reviews in Environmental Science and Technology 32: 17-72.

Perry, C.C. (2003). Silicification: the processes by which organisms capture and mineralize silica. In : Biomineralization. Vol. 54, pp. 291-324. Eds, P.M. Dove and J.J. De Yoreo. Mineralogical Society of America, Washington D.C.

Peters, R.W., Shem, L. (1995). Treatment of soils contaminated with heavy metals. In: Metal Speciation and Contamination of Soil, pp. 255-274. Eds, H.E. Allen, C.P. Huang, G.W. Bailey and A.R. Bowers. Lewis Publishers, Boca Raton, FL

Phillips, S.E., Taylor, M.L. (1976). Oxidation of arsenite to arsenate by *Alcaligenes faecalis*. Applied and Environmental Microbiology 32: 392-399.

Picardal, F.W., Arnold, R.G., Couch, H., Little, A.M., Smith, M.E. (1993). Involvement of cytochromes in the anaerobic biotransformation of tetrachloromethane by *Shewanella putrefaciens* 200. Applied and Environmental Microbiology 59: 3763-3770.

Pierce, M.L., Moore, C.B. (1982). Adsorption of arsenite and arsenate on amorphous iron hydroxide. Water Research 16: 1247-1253.

Postma, D., Boesen, C. (1991). Nitrate reduction in an unconfined sandy aquifer: Water chemistry, reduction processes, and geochemical modeling. Water Resources Research 27: 2027-2045.

Richard, F.C., Bourg, A.C.M. (1991). Aqueous geochemistry of chromium: a review. Water Research 25: 807-816.

Riley, R.G., Zachara, J.M., Wobber, F.J. (1992). U.S. Department of Energy, Washington, D.C.

Robinson, J.B., Tuovinen, O.H. (1984). Mechanism of microbial resistance and detoxification of mercury and organomercury compounds: Physiological, biochemical, and genetic analyses. Microbiology Reviews 48: 95-124.

Roden, E.E., Leonardo, M.R., Ferris, F.G. (2002). Immobilization of strontium during iron biomineralization coupled to dissimilatory hydrous ferric oxide reduction. Geochimica et Cosmochimica Acta 66: 2823-2839.

Rosen, B.P. (2002). Biochemistry of arsenic detoxification. FEBS Letters 52 9: 86-92.

Rosso, K.M., Zachara, J.M., Fredrickson, J.K., Gorby, Y.A., Smith, S.C. (2003). Nonlocal bacterial electron transfer to hematite surfaces. Geochimica et Cosmochimica Acta 67: 1081-1087.

Rothstein, A., Meier, R. (1951). The relationship of the cell surface to metabolism. VI. The chemical nature of uranium complexing groups of the cell surface. Journal of Cell. Comp. Physiology 38: 245-270.

Sakita, S., Kusuda, T. (2000). Modeling and simulation with microsites on vertical concentration profiles in sediments of aquatic zones. Water Science and Technology 42: 409-415.

Santini, J.M., Sly, L.I., Schnagl, R.D., Macy, J.M. (2000). A new chemolithoautotrophic arsenite-oxidizing bacterium isolated from a gold mine: Phylogenetic, physiological, and preliminary biochemical studies. Applied Environmental Microbiology 66: 92-97.

Sauvé, S. (2002). Speciation of metals in soils. In : Bioavailability of Metals in Terrestrial Ecosystems: Importance of Partitioning for Bioavailability to Invertebrates, Microbes, and Plants, pp. 158. Ed, H.E. Allen. SETAC Press, Pensacola, FL.

Schlautman, M.A., Han, I. (2001). Effects of pH and dissolved oxygen on the reduction of hexavalent chromium by dissolved ferrous iron in poorly buffered aqueous systems. Water Research 35: 1534-1546.

Schmieman, E.A., Petersen, J.N., Yonge, D.R., Johnstone, D.L., Bereded-Samuel, Y., Apel, W.A., Turick, C.E. (1997). Bacterial reduction of chromium. Applied Biochemistry and Biotechnology 63-65: 855-864.

Schwarzenbach, R.P., Stierli, R., Lanz, K., Zeyer, J. (1990). Quinone and iron porphyrin mediated reduction of nitroaromatic compounds in homogeneous aqueous solution. Environmental Science and Technology 24: 1566-1574.

Scott, D.T., McKnight, D.M., Blunt-Harris, E.L., Kolesar, S.E., Lovley, D.R. (1998). Quinone moieties act as electron acceptors in the reduction of humic substances by humics-reducing microorganisms. Environmental Science and Technology 32: 2984-2989.

Sedlak, D.L., Chan, P.G. (1997). Reduction of hexavalent chromium by ferrous iron. Geochimica et Cosmochimica Acta 61: 2185-2192.

Senesi, N., Miano, T.M. 1995. The role of abiotic interactions with humic substances on the environmental impact of organic pollutants. In : Environmental Impact of Soil Component Interactions, Vol. 1, pp. 311-335. Eds, P.M. Huang, J. Berthelin, J.-M. Bollag, W. B. McGill and A.L. Page. Lewis Publishers, Boca Raton.

Senko, J.M., Istok, J.D., Suflita, J.M., Krumholz, L.R. (2002). In-situ evidence for uranium immobilization and remobilization. Environmental Science and Technology 36: 1491-1496.

Shelobolina, E.S., Sullivan, S.A., O'Neill, K.R., Nevin, K.P., Lovley, D.R. (2004). Isolation, characterization, and U(VI)-reducing potential of a facultatively anaerobic, acid-resistant bacterium from low-pH, nitrate- and U(VI)-contaminated subsurface sediment and description of Salmonella subterranea sp. nov. Applied and Environmental Microbiology 70: 2959-2965.

Smith, A.H., Lingas, E.O., Rahman, M. (2000). Contamination of drinking water by arsenic in Bangladesh: a public health emergency. Bull. WHO 78: 1093-1103.

Smith, K.A. (1990). Anaerobic zones and denitrification in soil: modelling and measurement. In : Denitrification in Soil and Sediment, pp. 229-244.Eds, N.P. Revsbech and J. Sørensen Plenum Press, New York.

Smith, T., Pitts, K., McGarvey, J.A., Summers, A.O. (1998). Bacterial oxidation of mercury metal vapor, Hg(0). Applied and Environmental Microbiology 64: 1328-1332.

Sobolev, D., Roden, E.E. (2002). Evidence for rapid microscale bacterial redox cycling of iron in circumneutral environments. Antonie von Leeuwenhoek 81: 587-597.

Sobolev, D., Roden, E.E. (2004). Characterization of a neutrophilic, chemolithoautotrophic Fe(II)-oxidizing ß-proteobacterium from freshwater sediments. Geomicrobiology Journal 21: 1-10.

Sokolov, I., Smith, D.S., Henderson, G.S., Gorby, Y.A., Ferris, F.G. (2001). Cell surface electrochemical heterogeneity of the Fe(III)-reducing bacteria *Shewanella putrefaciens.* Environmental Science and Technology 35: 341-347.

Spring, S. (2004). The genera *Leptothrix* and *Sphaerotilus.* In : The Prokaryotes: An Evolving Electronic Resource for the Microbiological Community, 3rd edition, Release 3.9, 4/01/02; Ed, M. Dworkin Springer-Verlag, New York. http://link.springer-ny.com/link/service/books/10125/.

Stocks-Fischer, S., Galinat, J.K., Bang, S.S. (1999). Microbiological precipitation of CaCO$_3$. Soil Biology and Biochemistry 31: 1563-1571.

Stolz, J., Ellis, D., Switzer Blum, J., Ahmann, D., Lovley, D., Oremland, R. (1999). Sulfurospirillum barnesii sp. nov. and Sulfurospirillum arsenophilum sp. nov., new members of the Sulfurospirillum clade of the epsilon Proteobacteria. International Journal of Systematic Bacteriology 49: 1177-1180.

Stolz, J.F., Oremland, R.S. (1999). Bacterial respiration of arsenic and selenium. FEMS Microbiology Reviews 23: 615-627.

Straub, K.L., Benz, M., Schink, B., Widdel, F. (1996). Anaerobic, nitrate-dependent microbial oxidation of ferrous iron. Applied and Environmental Microbiology 62: 1458-1460.

Straub, K.L., Buchholz-Cleven, B.E.E. (1998). Enumeration and detection of anaerobic ferrous iron-oxidizing, nitrate-reducing bacteria from diverse European sediments. Applied and Environmental Microbiology 64: 4846-4856.

Suarez, P., Reyes, R. (2002). Heavy metal incorporation in bacteria and its environmental significance. Interciencia 27: 160-165.

Sunda, W.G., Huntsman, S.A. (1998). Processes regulating cellular metal accumulation and physiological effects: Phytoplankton as model systems. Science of the Total Environment 219: 165-181.

Takahashi, Y., Minai, Y., Ambe, S., Makide, Y., Ambe, F. (1999). Comparison of adsorption behavior of multiple inorganic ions on kaolinite and silica in the presence of humic acid using the multitracer technique. Geochimica et Cosmochimica Acta 63: 815-836.

Taunton, A.E., Welch, S.A., Banfield, J.F. (2000). Microbial controls on phosphate and lanthanide distributions during granite weathering and soil formation. Chemical Geology 169: 371-382.

Tebo, B.M., Obraztsova, A.Y. (1998). Sulfate-reducing bacterium grows with Cr(VI), U(VI), Mn(IV), and Fe(III) as electron acceptors. FEMS Microbiology Letters 162: 193-198.

Tesoriero, A.J., Pankow, J.F. (1996). Solid solution partitioning of Sr^{2+}, Ba^{2+}, and Cd^{2+} to calcite. Geochimica et Cosmochimica Acta 60: 1053-1063.

Tilton, R.C., Gunner, H.B., Litsky, W. (1967). Physiology of selenite reduction by *Enterococci* I. Influence of environmental variables. Canadian Journal of Microbiology 13: 1175-1182.

Tokunaga, T.K., Wan, J., Firestone, M.K., Hazen, T., Olson, K.R., Herman, D.J., Sutton, S.R., Lanzirotti, A. (2003). In situ reduction of chromium (VI) in heavily contaminated soils through organic carbon amendment. Journal of Environmental Quality 32: 1641-1649.

Tomei, F.A., Barton, L.L., Lemanski, C.L., Zocco, T.G., Fink, N.H., Sillerud, L.O. (1995). Transformation of selenate and selenite to elemental selenium by *Desulfovibrio desulfuricans*. Journal of Industrial Microbiology 14: 329-336.

Tsapin, A.I., Vandenberghe, I., Nealson, K.H., Scott, J.H., Meyer, T.E., Cusanovich, M.A., Harada, E., Kaizu, T., Akutsu, H., Leys, D., Van Beeumen, J.J. (2001). Identification of a small tetraheme cytochrome c and a flavocytochrome c as two of the principal soluble cytochromes c in Shewanella oneidensis strain MR1. Applied and Environmental Microbiology 67: 3236-3244.

Turick, C.E., Graves, C., Apel, W.A. (1998). Bioremediation potential of Cr(VI)-contaminated soil using indigenous microorganisms. Bioremediation Journal 2: 1-6.

Turick, C.E., Tisa, L.S., Caccavo, F., Jr. (2002). Melanin production and use as a soluble electron shuttle for Fe(III) oxide reduction and as a terminal electron acceptor by Shewanella algae BrY. Applied and Environmental Microbiology 68: 2436-2444.

Urrutia-Mera, M., Beveridge, T.J. (1993). Mechanism of silicate binding to the bacterial cell wall in bacillus subtilis. Journal of Bacteriology 175: 1936-1945.

vanden Hoven, R.N., Santini, J.M. (2004). Arsenite oxidation by the heterotroph Hydrogenophaga sp. str. NT-14: the arsenite oxidase and its physiological electron acceptor. Biochimica et Biophysica Acta (BBA) - Bioenergetics 1656: 148-155.

Vaughan, D.J., Lennie, A.R. (1991). The iron sulfide minerals - their chemistry and role in nature. Science Progress 75: 371-388.

Veizer, J. (1990). Trace elements and isotopes in sedimentary carbonates. In : Carbonates: Mineralogy and Chemistry, Vol. 11, pp. 399. Ed, R.J. Reeder. Mineralogical Society of America, Washington D.C.

Verrecchia, E.P. (1994). Microbial and surficial origin of the calcrete laminar horizon. Bulletin De La Societe Geologique De France 165: 583-592.

Viamajala, S., Peyton, B.M., Petersen, J.N. (2003). Modeling chromate reduction in *Shewanella oneidensis* MR-1: Development of a novel dual-enzyme kinetic model. Biotechnology and Bioengineering 83: 790-797.

Wang, Y.F., Nahon, D., Merino, E. (1994). Dynamic-model of the genesis of calcretes replacing silicate rocks in semiarid regions. Geochimica et Cosmochimica Acta 58: 5131-5145.

Wang, Y.-T. (2000). Microbial reduction of chromate. In :Environmental Microbe-Metal Interactions, pp. 225-235. Ed, D.R. Lovley. ASM Press, Washington, D.C.

Wang, Y.-T., Shen, H. (1995). Bacterial reduction of hexavalent chromium. Journal of Industrial Microbiology 14: 159-163.

Wang, Y.-T., Xiao, C.S. (1995). Factors affecting hexavalent chromium reduction in pure cultures of bacteria. Water Research 29: 2467-2474.

Warren, L.A., Maurice, P.A., Parmar, N., Ferris, F.G. (2001). Microbially mediated calcium carbonate precipitation: Implications for interpreting calcite precipitation and for solid-phase capture of inorganic contaminants. Geomicrobiology Journal 18: 93-115.

Weber, K.A., Picardal, F.W., Roden, E.E. (2001). Microbially catalyzed nitrate-dependent oxidation of biogenic solid-phase Fe(II) compounds. Environmental Science and Technology 35: 1644-1650.

Welch, S.A., Taunton, A.E., Banfield, J.F. (2002). Effect of microorganisms and microbial metabolites on apatite dissolution. Geomicrobiology Journal 19: 343-367.

White, A.F., Peterson, M.L. (1996). Reduction of aqueous transition metal species on the surfaces of Fe(II)-containing oxides. Geochimica et Cosmochima Acta 60: 3799-3814.

Widdel, F., Schnell, S., Heising, S., Ehrenreich, A., Assmus, B., Schink, B. (1993). Ferrous iron oxidation by anoxygenic phototrophic bacteria. Nature 362: 834-836.

Wielinga, B., Mizuba, M.M., Hansel, C.M., Fendorf, S. (2001). Iron promoted reduction of chromate by dissimilatory iron-reducing bacteria. Environmental Science and Technology 35: 522-527.

Wildung, R.E., Gorby, Y.A., Krupka, K.M., Hess, N.J., Li, S.W., Plymale, A.E., McKinley, J.P., Fredrickson, J.K. (2000). Effect of electon donor and solution chemistry on products of dissimilatory reduction of technetium by *Shewanella putrefaciens*. Applied and Environmental Microbiology 66: 2451-2460.

Wu, J., Roth, C.B., Low, P.F. (1988). Biological reduction of structural iron in sodium-nontronite. Soil Science Society of America Journal 52: 295-296.

Xing, G.X., Xu, L.Y., Hou, W.H. (1995). Role of amorphous Fe oxides in controlling retention of heavy metal elements in soils. In : Environmental Impact of Soil Component Interactions, Vol. 2, pp. 63-74. Eds, P.M. Huang, J. Berthelin, J.-M, Bollag, W.B. McGill and A.L. Page. Lewis Publishers, Boca Raton.

Yang, J., Mosby, D.E., Casteel, S.W., Blanchar, R.W. (2001). Lead immobilization using phosphoric acid in a smelter-contaminated urban soil. Environmental Science and Technology 35: 3553-3559.

Yee, N., Fein, J. (2001). Cd adsorption onto bacterial surfaces: A universal adsorption edge? Geochimica et Cosmochimica Acta 65: 2037-2042.

Yee, N., Fein, J.B. (2002). Does metal adsorption onto bacterial surfaces inhibit or enhance aqueous metal transport? Column and batch reactor experiments on Cd-*Bacillus subtilis*-quartz systems. Chemical Geology 185: 303-319.

Yee, N., Fein, J.B. (2003). Quantifying metal adsorption onto bacteria mixtures: a test and application of the surface complexation model. Geomicrobiology Journal 20: 43–60.

Zachara, J.M., Fredrickson, J.K., Li, S.-M., Kennedy, D.W., Smith, S.C., Gassman, P.L. (1998). Bacterial reduction of crystalline Fe^{3+} oxides in single phase suspensions and subsurface materials. American Mineralogist 83: 1426-1443.

Zachara, J.M., Fredrickson, J.K., Smith, S.C., Gassman, P.L. (2001). Solubilization of Fe(III) oxide-bound trace metals by a dissimilatory Fe(III) reducing bacterium. Geochimica et Cosmochima Acta 65: 75-93.

Zachara, J.M., McKinley, J.P. (1993). Influence of hydrolysis on the sorption of metal-cations by smectites - importance of edge coordination reactions. Aquatic Sciences 55: 250-261.

Zachara, J.M., Smith, S.C. (1994). Edge complexation reactions of cadmium on specimen and soil- derived smectite. Soil Science Society of America Journal 58: 762-769.

Zachara, J.M., Smith, S.C., Liu, C.X., McKinley, J.P., Serne, R.J., Gassman, P.L. (2002). Sorption of Cs^+ to micaceous subsurface sediments from the Hanford site, USA. Geochimica et Cosmochimica Acta 66: 193-211.

Zobrist, J., Dowdle, P.R., Davis, J.A., Oremland, R.S. (2000). Mobilization of arsenite by dissimilatory reduction of adsorbed arsenate. Environmental Science and Technology 34: 4747-4753.

CHAPTER 4

Influence of Long-Term Application of Treated Oil Refinery Effluent on Soil Health

Iqbal Ahmad[1], S. Hayat[2], A. Ahmad[2],
A. Inam[2] and Samiullah[2]

[1]Department of Agricultural Microbiology,
Aligarh Muslim University, Aligarh-202 002, INDIA
[2]Department of Botany, Aligarh Muslim University,
Aligarh-202 002, INDIA.

The chemical characteristics, heavy metal content and microbiological characteristics of soil irrigated with effluent from the Mathura Oil Refinery from 1987 to 2001, were compared with soil receiving fresh water. This field was used to cultivate various seasonal crops year-round. In the last 15 years, the above soil characteristics did not significantly change in response to effluent application. However, a non-significant accumulation of certain heavy metals in the soil was observed. Similarly, microbial populations and diversity did not significantly shift. It may, therefore, be suggested that properly treated oil refinery effluent may be safely used as an irrigant to agricultural fields. However, monitoring of soil health status is necessary for long-term application.

INTRODUCTION

Fresh water and groundwater resources are becoming scarce to farmers, not only in India but in many developed countries. This, therefore, leaves few options for farmers but to grow crops either under rain-fed conditions and/or to use municipal and industrial treated wastewater. Moreover, the use of industrial effluents in agricultural fields may be the most economical outlet for effluents that also provides an opportunity to recycle beneficial plant nutrients and organic matter to the soil and plants for optimal production. However, the quality of the effluent as an irrigant, even from the same source, fluctuates regularly, therefore requiring proper monitoring and treatment before being made available to farmers.

Many industrial effluents such as those from paper, sugar, brewery and textile industries are known to improve the performance of various crops without adversely affecting soil properties (Boll and Bell, 1978; Ajmal and Khan, 1983; 1984; 1985; Samuel, 1986; Devarajan *et al.*, 1994). Sugar distillery effluents contain many essential plant nutrients and are considered as liquid manure but only limited applications were required (Samuel, 1986; Devarajan *et al.*, 1994). Similarly, the effluent from Mathura Oil Refinery, a constituent of Indian Oil Corporation Ltd., India was found to be suitable for the cultivation of various short- and long-duration crops under field conditions (Aziz *et al.*, 1996; Hayat *et al.*, 2002a; Ahmad *et al.*, 2003). However, these effluents contain varying quantities of potentially toxic metals. Repeated application of this wastewater may, therefore, affect soil microbial populations and associated activities which are key factors regulating the cycling of elements and the maintenance of soil fertility.

This review covers the suitability of oil refinery effluent as an irrigant and also the response of the soil in terms of chemical and microbial characteristics and accumulation of heavy metals, to its long-term application.

THE OIL REFINERY

The Mathura Oil Refinery, an establishment of Indian Oil Corporation Ltd., New Delhi, India, processes indigenous Bombay high crude and various imported crudes. The capacity of the refinery is 8.0 MMTPA (million metric ton per annum). The water requirement for the refinery is met from Koyala and Keetham Lakes near Mathura. The refinery discharges about 16,000 m^3 effluent per day. This effluent, before being discharged into Brari Minor Irrigation Canal and Yamuna River, is treated through an ETP (effluent treatment plant), where physical, chemical and biological methods are employed to remove various contaminants. The two main systems employed in effluent treatment include : (a) plant pre-treatment (sour water stripping and neutralization) and (b) ETP, which involves physical treatment (oil/water separation), chemical treatment (removal of sulphides) and biological treatment (biodegradation of organics, removal of settleable solids and natural aeration in polishing ponds). The oily effluent is then subjected to physical treatment for oil removal in a separator. Spent caustic is subjected to chemical treatment with enhanced oxidation using H_2O_2.

The combined stream is subjected to biological treatment in a trickling filter followed with an activated sludge process (ASP) in an aeration tank where refinery township domestic sewage is also mixed. The treated stream from the ASP is subjected to sedimentation in a clarifier. It next undergoes polishing in ponds through natural aeration and symbiotic processes by underwater plants and microorganisms. The effluent so released from the polishing ponds meets the MINAS (Minimum National Standards), the prescribed effluent discharge standards recommended for oil refineries. A part of this treated effluent is recycled for various uses within the refinery, resulting in the conservation of fresh water. The remaining quantity of the effluent is released into the Brari Minor Irrigation Canal and the Yamuna River. Local farmers use the effluent as an irrigant. This group of scientists from Aligarh Muslim University initiated a long-term field study (1987-2001) to evaluate various aspects of soil health and crop productivity by irrigating the field with treated effluent.

EXPERIMENTAL SITE AND CROPPING PATTERN

The Mathura Oil Refinery is situated on National highway No. 2 and is about 15 km from Mathura City towards Agra, U.P., India (latitude 27°22′N, longitude 77°40′E and altitude 174 M above MSL). Agricultural farm was established in 1987 in the vicinity of the oil refinery and maintained by Aligarh Muslim University, Aligarh up to 2002. Identical plots were prepared in a way where field experiments could be conducted in a split-plot design. The farm was kept under cultivation by growing a number of crops, i.e., wheat (*Triticum aestivum*), mustard (*Brassica juncea*), berseem (*Trifolium alexandrium*), lentil (*Lens culinaris*), moong (*Vigna radiata*), arhar (*Cajanus cajan*), sugarcane (*Saccharum officinarum*), sugarbeet (*Beta vulgaris*), radish (*Raphanus sativus*) and carrot (*Daucus carrota*) in their respective season (Table 4.1).

IRRIGATIONAL QUALITY OF FRESH WATER AND THAT OF THE TREATED EFFLUENT

The quality of fresh water (FW) and of treated effluent (TE) was evaluated for quality parameters recommended for agricultural use by the Food and Agriculture Organization of the United Nations (1985, 1994) during the entire course of study (1987-2001). Standard methods were used in studying various chemical characteristics (Ghosh *et al.*, 1983) and heavy metal content (Vanloon and Lichwa, 1973).

The pH values for both irrigants are similar but TE had higher electrical conductivity (EC) and chemical oxygen demand (COD) compared with FW. Similarly, the level of important inorganic ions essential for plant growth (nitrate, phosphate, potassium, calcium, magnesium and sulphate) was higher in TE than FW. Moreover, the level of sodium and chloride were under the permissible limit as per Indian Standards (Gupta, 1988); otherwise, they may have caused toxicity to plants. A total of 22 samples were analysed to assess the heavy metal (Cd, Ni, Pb, Zn, Cr, Fe, Mn, Cu) status in the irrigants (Table 4.2). Concentrations of heavy metals were higher in TE than FW; however, the level of these elements was under the

Table 4.1   Crops and their year of cultivation (1987–2001)

Wheat	Mustard	Berseem	Lentil	Moong	Arhar	Sugarcane	Sugarbeet	Radish	Carrot
1987–88	1993–94	1988–89	1991–92	1988	1998	1998–99	1998–99	1998–99	1998–99
1988–89	1994–95	1989–90	1993–94	1989	1999	1999–2000	1999–2000	1999–2000	1999–2000
1989–90	1996–97		1994–95	1990	2000	2000–2001	2000–2001	2000–2001	2000–2001
1990–91	1998–99			1992	2001				
1996–97				1995					
1997–98				1996					
				1997					
				1998					
				1999					

Table 4.2 Irrigational water quality (All determinations in mg l^{-1} or as specified; Range values during 15 year)

Characteristics	Fresh water	Treated effluent
PH	7.5–8.0	7.7–8.4
EC (dS m^{-1})	0.56–1.10	0.88–1.39
COD	24.1–40.4	46.8–80.3
BOD	3.8–4.1	4.2–4.7
TDS	560–820	705–990
Bicarbonate	126–203	91–228
Carbonate	N.D.	11.0–14.0
Chloride	66.0–160.0	143.0–181.0
Sulphate	41.3–95.0	65.0–123.3
Nitrate	0.26–0.78	0.30–2.04
Sodium	24.0–70.0	55.3–74.8
Potassium	6.0–13.0	13.0–20.0
Magnesium	13.0–60.0	21.0–66.0
Calcium	14.8–43.0	23.0–60.0
Oil & grease	N.D.	0.8–1.3
Cadmium (ppm)	N.D.	0.012–0.104
Nickel (ppm)	0.25–0.31	0.36–0.41
Lead (ppm)	0.01–0.02	0.02–0.03
Zinc (ppm)	0.11–0.12	0.25–0.30
Chromium (ppm)	0.02–0.03	0.04–0.045
Iron (ppm)	0.75–0.90	0.93–1.15
Manganese (ppm)	0.02–0.03	0.05–0.06
Copper (ppm)	0.025–0.31	0.04–0.05

permissible limits, as recommended by Krishna Murti and Vishwanathan, (1991).

CROP YIELD IN RELATION TO TE AND FW

The reported study is the first time that the effluent from an oil refinery was used to cultivate various crops with a control, during 1987-2001, to evaluate their performance.

The mean values for yield are presented in Table 4.3. Yield of all crops irrigated with TE increased, compared with those receiving FW. This beneficial effect of TE may be attributed to the presence of additional essential nutrients in TE than FW (Table 4.3). This result gets additional support from the observed increase in plant biomass, irrigated with TE (Aziz et al., 1994, 1996; Hayat et al., 2000, 2002a) under higher soil nutrient status (Marchner, 2003). Our findings are in agreement with those of other

Table 4.3 The biological yield (q ha⁻¹) of the crops receiving fresh water (FW) or treated effluent (TE) grown during 1987-2001

Wheat

Crop	1987-88		1988-89		1989-90		1990-91		1996-97		1997-98	
	FW	TE	FW	TE	FW	TE	FW	TE	FW	TE	FW	TE
Wheat	49.1	51.7	46.4	50.6	44.8	48.8	51.04	55.21	41.95	42.87	41.80	42.83

Mustard

Crop	1993-94		1994-95		1996-97		1998-99	
	FW	TE	FW	TE	FW	TE	FW	TE
Mustard	10.16	12.08	10.88	12.96	9.40	10.20	9.37	10.20

Berseem

Sampling	1988-89		1989-90	
	FW	TE	FW	TE
I Cut	204.4	222.4	225.5	250.0
II Cut	277.1	326.7	232.6	281.2
III Cut	260.1	274.6	266.5	283.3
IV Cut	351.2	363.5	274.2	286.8

Lentil

Crop	1991-92		1993-94		1994-95	
	FW	TE	FW	TE	FW	TE
Lentil	13.31	14.08	13.64	15.70	15.53	17.64

Moong

Crop	1988		1989		1990		1992		1995		1996		1997		1998		1999	
	FW	TE	FW	TE	FW	TE	FW	TE	FW	TE	FW	TE	FW	TE	FW	TE	FW	TE
Moong	13.2	12.2	11.9	10.0	11.24	9.73	12.32	10.81	12.88	11.25	14.05	12.73	10.09	10.0	9.58	9.67	9.94	10.32

Arhar

Crop	1998		1999		2000		2001	
	FW	TE	FW	TE	FW	TE	FW	TE
Arhar	11.95	14.05	10.80	12.50	11.30	13.40	13.35	14.99

Sugarcane, Sugarbeet, Radish, Carrot

Crop	1998-99		1999-2000		2000-2001	
	FW	TE	FW	TE	FW	TE
Sugarcane	98.4	94.6	97.28	102.99	103.68	109.34
Sugarbeet	175.45	189.30	188.31	196.45	189.22	198.10
Radish	173.10	180.25	169.45	177.32	170.32	179.40
Carrot	165.66	173.45	177.11	186.89	178.71	186.15

researchers on paper factory effluent (Rajannan and Oblisami, 1979), sugar mill effluent (Day et al., 1979), agricultural effluent (Ajmal and Khan, 1983) and thermal power effluent (Inam et al., 2002). However, the treated effluents from different sources were not similar in their characteristics.

SOIL HEALTH

Maintenance of soil health is of prime importance in sustainable agriculture. Parameters that reflect soil health include nutrient status and chemical and microbiological characteristics as well as soil enzyme status. All these characteristics plus the cation exchange capacity of the soil, irrigated with the effluents from agricultural industry and the sewage waste, improves from the addition of organic matter (Ajmal and Khan, 1983; Smith, 1996).

The quality of the effluents released from different industries fluctuate markedly and may not improve soil health; therefore, effluent quality must be continuously monitored and certified from standard agencies before being made available to farmers as an irrigant. We have been monitoring the health status of soil receiving refinery effluents, of which some important parameters follow.

Chemical Characteristics of Soil

The quality of the effluent released from Mathura Oil Refinery was regularly monitored and the level of important soil characteristics (e.g. pH, electrical conductivity, cation exchange capacity, total dissolved solids, chloride, bicarbonate, potassium, sodium, calcium, magnesium, nitrate and phosphorus) observed at the initial stages of the project and after 15 years, during which the soil was irrigated with FW or TE, are shown in Table 4.4. In contrast to other published literature the present study resulted in no significant change in soil chemical characteristics (Table 4.4). Initially, soil pH was slightly alkaline (Table 4.4) but started decreasing linearly over time as the soil was irrigated with FW or TE. However, the electrical conductivity exhibited a reverse trend. Moreover, the soil salt level was maintained at permissible levels probably due to consistency in the concentration of the salt in the treated effluent. A completely different picture emerges if one observes the chemical characteristics and fertility of soil receiving effluents from other industries (Rao et al., 1993; Kansal et al., 1993).

Heavy Metal Content in Soil

Even though the effluent passes through various phases of purification it is still loaded with relatively higher quantities of heavy metals (Cd, Ni, Pb, Zn, Cr, Fe, Mn and Cu), compared with fresh water (Table 4.5). The soil irrigated

Table 4.4   Physico-chemical characteristics of soil irrigated with fresh water (FW) and treated effluent (TE) (All the determinations in mg l^{-1} in 1:5 (soil: water) extract or as specified)

Characteristics	1987	After 15 years of irrigation	
		FW	TE
pH[a]	8.4	7.9–8.1	8.2–8.4
E.C[a] (dsm^{-1})	0.45	0.85–0.93	1.2–1.3
Chloride	44.0	24.0–33.0	23.0–42.0
Bicarbonate	192.0	195.0–247.0	210–283.0
Potassium	12.0	13.0–24.0	14.0–26.0
Sodium	27.0	29.0–45.0	31.0–65.0
Calcium	21.0	24.0–37.0	22.0–38.0
Magnesium	20.0	23.0–31.0	22.0–48.0
Nitrate	0.20	0.27–0.31	0.26–0.33
Phosphate	0.24	0.30–0.35	0.35–0.48
Total dissolved solids	618.0	720–835	725–905

Table 4.5   Heavy metal content (μg g^{-1}) in soil, irrigated with fresh water (FW) or treated effluent (TE) (Range values during 15 years)

Metals	FW	TE
Cd	0.70–1.25	N.D.–1.36
Ni	22.0–28.5	30.0–34.0
Pb	24.8–27.5	30.7–57.8
Zn	40.2–49.2	41.6–58.3
Cr	23.2–29.9	32.7–34.3
Fe	110.1–114.4	115.2–131.6
Mn	0.92–1.02	1.05–1.13
Cu	9.7–16.4	9.4–23.2

N.D. = Non detectable

with TE would therefore have been enriched with TE elements from each irrigation cycle during 1987-2001 (Table 4.5). However, the concentrations for all heavy metals in the soil are within permissible limits and show no significant correlation with those of TE or FW. The data clearly demonstrates the dependence of the fate of each metal added to the soil complex on various factors that includes their interaction with plants (Juste and Mench, 1992) and cropping pattern.

Microbiological Characteristics of Soil

A shift in microbial characteristics is expected to develop because of the composition of the effluent, in particular the metals content. The presence of heavy metals in sewage sludge influenced the microbes and the associated

Table 4.5a The maximum limits of metals, in irrigation water, soil and crop, above which they become toxic

Metals	Irrigation water* (mg l^{-1})	Soil** (mg kg^{-1})
Cd	0.01	1.62
Ni	0.20	33.7
Pb	5.0	29.2
Zn	2.0	59.8
Cr	0.10	84.0
Fe	5.0	32000.0
Mn	0.20	760.0
Cu	0.20	25.8

*Gupta, I.C. (1988)
**Smith, S.R. (1996).

activities in soil including microbial biomass, soil ATP concentration, dehydrogenase activity, N_2 fixation by heterotrophs, nitrification of added nitrate nitrogen, and nitrogenase activity (Brookes and McGrath, 1984; McGrath et al., 1988). However, certain characteristics like phosphatase activity, mineralization of native soil organic carbon and nitrogen and adenylate energy change were less sensitive to the toxic effects of heavy metals (Brookes and McGrath, 1987).

Studies in relation to oil refinery effluent on microbial process in soil are lacking. The microbiological characteristics in this soil were studied between 1997-2001. The studies were largely confined to microbial diversity of the soil, in terms of major microbial groups which could easily be cultivated on nutrient agar plates (heterotrophic aerobic bacteria, actinomycetes, asymbiotic nitrogen fixers and filamentous fungi). Moreover, nitrifying bacteria and cellulolytic microorganisms were also detected (Table 4.6). Plate counts of these groups of soil microorganisms are comparable with that of typical soil microflora reported by Subba Rao (1977) and Alexander (1984). Viable counts of various groups of microorganisms in the soil samples, irrigated with either FW or TE, showed no significant difference (Table 4.6 and Hayat et al., 2002b). The data gives the impression that the dynamics of microbial populations in the soil irrigated for a long duration with oil refinery effluent is maintained. Possible reasons for these findings are: (a) the increase in heavy metal content is not to a significant level and is therefore non-toxic; (b) microbes mutate to acquire resistance to the increased level of the metals, and (c) the complex nature of soil-metal microbe interactions with these changes, at the level of the microbes, might be overcoming the toxic effect of the metals.

It is obvious from these results that no apparent adverse effect on soil health at the agricultural farm at the Mathura Oil Refinery is exhibited, which might be due to the maintenance of effluent quality.

Table 4.6 Microbial diversity of soil, receiving fresh water (FW) or treated effluent (TE) during 1997-2001

Microbial diversity	Viable count (CFU/g of soil)									
	1997		1998		1999		2000		2001	
	FW	TE	FW	TE	FW	TE	FW	TE	FW	TE
Total aerobic heterotrophic bacteria	2.6×10^6	7.0×10^6	2.5×10^6	7.5×10^6	7.2×10^7	6.2×10^7	2.4×10^6	6.9×10^6	7.6×10^7	5.9×10^7
Total actinomycetes	1.2×10^5	2.0×10^4	2.4×10^5	2.5×10^4	2.8×10^5	1.7×10^5	1.3×10^5	2.7×10^4	3.2×10^5	2.6×10^5
Total fungi	7.5×10^5	2.5×10^4	6.8×10^5	3.2×10^4	2.1×10^5	2.8×10^5	6.6×10^5	2.9×10^4	3.2×10^4	3.6×10^5
Total asymbiotic nitrogen fixer	1.2×10^5	5.5×10^4	2.2×10^5	5.0×10^4	6.8×10^5	2.5×10^5	1.3×10^5	6.3×10^4	6.4×10^5	3.3×10^5
Cellulolytic bacteria	-	-	-	-	-	-	++	+	+++	+++
Nitrifying bacteria	-	-	-	-	-	-	++	+	+++	+++

Values are mean of 22 samples.

+ = Present in less than 25% samples

++ = Present in more than 50% samples

+++ = Present in more than 75% samples

- = Not done

Metal Tolerance Levels of Soil Bacteria

The interaction effect of heavy metals and soil microorganisms is quite complex; therefore, published reports are often contradictory. Such an observation has been expressed by Smith (1991) in his studies on the impact of sludge application to soil. However, the bioavailability of heavy metals and their involvement as toxicants to microorganisms is very much determined by differences in pH, status of organic matter and chemical properties of the soil. Moreover, the degree of adaptability of soil organisms to the altered conditions may be an additional source of variability in the observations (Tyler, 1981; Olson and Thornton, 1982; Baath, 1989; Ansari et al., 2001; Ahmad et al., 2001).

The goal of the present study was to concentrate on the possible frequency of occurrence of metal-tolerant microbial populations in the soil irrigated with oil refinery effluent or fresh water. The tolerance of these soil organisms to heavy metals was determined by viable counts on requisite medium plates supplemented with 200 or $400\,\mu g\,ml^{-1}$ of heavy metals (Ni, Cd, Pb, Co, Cu, Cr, Zn and Hg). At 200 $\mu g\,ml^{-1}$ concentration no significant inhibition of bacterial growth was observed. Thus the majority of the bacterial populations, in each group, (aerobic heterotrophs, asymbiotic nitrogen fixers and actinomycetes) tolerated $200\,\mu g\,ml^{-1}$ concentrations of most of the heavy metals. However, a significant decrease in viable counts was observed against Cd and Hg as compared with the control (i.e. without metal) (Table 4.7). At the higher metal concentration $(400\,\mu g\,ml^{-1})$ the growth of aerobic heterotrophs was observed on the plates supplemented with Pb, Cr and Zn. Similarly, the growth of asymbiotic nitrogen fixers was evident on the medium treated with Ni, Cd, Pb, Cu and Zn. However, the growth of actinomycetes was completely inhibited at this higher concentration. These observations lead us to conclude that actinomycetes (Gram-positive bacteria) possessed a lower tolerance to the higher level $(400\,\mu g\,ml^{-1})$ of heavy metals than Gram-negative bacteria (asymbiotic nitrogen fixers). This behavioural difference to metals by Gram-positive and Gram-negative bacteria is possibly an expression of the variation in cell wall chemical composition (Hughes and Poole, 1989) and the genetic variability of the bacteria affecting their interaction with the metals (Hughes and Poole, 1989).

CONCLUSIONS

The following points emerge from this long-term study conducted at the agricultural farm of the Mathura Oil Refinery with various crops raised under fresh water or treated effluent:

1. The economic yield of the crops berseem, wheat, mustard, lentil, arhar, sugarbeet, radish, carrot, moong and sugarcane, on being irrigated

Table 4.7 Frequency of metal tolerant bacterial population in soil (Mean of 22 samples)

Heavy metals	Aerobic heterotrophic (CFU g⁻¹ of soil)			Potentially asymbiotic nitrogen fixers (CFU g⁻¹ of soil)			Actinomycetes (CFU g⁻¹ of soil)		
	Metal concentration (mg/ml)			Metal concentration (mg/ml)			Metal concentration (mg/ml)		
	Control	200	400	Control	200	400	Control	200	400
Ni	3.45×10^5	4.1×10^4	Nil	2.8×10^5	1.65×10^5	3.3×10^5	1.8×10^5	1.9×10^4	Nil
Cd	3.45×10^5	4.8×10^4	Nil	2.8×10^5	6.2×10^4	1.1×10^4	1.8×10^5	1.4×10^4	Nil
Pb	3.45×10^5	1.6×10^4	1.5×10^4	2.8×10^5	7.5×10^4	4.8×10^3	1.8×10^5	2.5×10^4	Nil
Co	3.45×10^5	1.1×10^5	Nil	2.8×10^5	1.3×10^5	Nil	1.8×10^5	4.3×10^4	Nil
Cu	3.45×10^5	6.4×10^3	Nil	2.8×10^5	1.0×10^5	1.2×10^4	1.8×10^5	6.4×10^4	Nil
Cr	3.45×10^5	1.3×10^5	1.4×10^4	2.8×10^5	3.8×10^4	Nil	1.8×10^5	2.0×10^4	Nil
Zn	3.45×10^5	2.1×10^5	6.4×10^4	2.8×10^5	1.65×10^5	5.3×10^4	1.8×10^5	3.8×10^4	Nil
Hg	3.45×10^5	2.5×10^2	Nil	2.8×10^5	3.4×10^4	Nil	1.8×10^5	1.9×10^4	Nil

Nil = At dilution 10^{-4}, 10^{-5}, 10^{-6} of soil.

with treated effluent increased compared with those irrigated with fresh water.

2. The soil health, in terms of its chemical characteristics and the status of heavy metals (Cd, Ni, Pb, Zn, Cr, Fe, Mn and Cu) is suitable for cultivation. All metal levels attained during this study period were below permissible limits.

3. As a natural phenomenon, microbial interactions with the soil maintain its fertility. In our case, microbial characteristics did not show any significant change in response to the effluent; therefore, the soil environment is conducive for normal growth of plants.

4. The majority of the soil microbial populations tolerated existing level of metals, indicating their intrinsic potential to survive in limited metal-contaminated soil.

5. Based on the chemical and microbiological studies of the treated effluent, it may not be premature to certify that Mathura Oil Refinery is managing its effluent quality so that it may be used by farmers as an irrigant.

FUTURE DIRECTIVES

Application of treated oil refinery effluent over the previous 15 years did not significantly impact crop yield and soil health, and concentrations of soil metals were maintained at a level below permissible limits. The agricultural fields are under continuous cultivation, with two-thirds of the crops rotated over the year, and metals continue to be added to the soil with each irrigation cycle. However, a slow increase in the heavy metal content of soil is recorded during the period of investigation, which highlights the importance of continuous monitoring of soil health using more sensitive parameters. Therefore, the following remedial measures may be suggested :

(a) Some sensitive indicators of soil health (e.g. soil enzyme activity, specific microbial activity and total microbial biomass) along with regular monitoring of the quality of the effluent, bioavailability of metal in soil and the leaching of the metals should be adopted.

(b) Farmers should be advised to not regularly irrigate fields solely with effluent but intermix with fresh water, which may be helpful to overcome any adverse effect due to long-term application of effluent.

REFERENCES

Ahmad, A., Inam, A., Ahmad, I., Hayat, S., Azam, Z.M., Samiullah (2003). Response of sugarcane to treated wastewater of oil refinery. Journal of Environmental Biology 24 : 141-146.

Ahmad, I., Hayat, S., Ahmad, A., Inam, A., Samiullah (2001). Metal and antibiotic resistance traits in *Bradyrhizobium* sp. (cajanus) isolated from soil receiving oil refinery wastewater. World Journal of Microbiology & Biotechnology 17: 379-384.

Ahmad, S., Yadav, J.N.S. (1988). Infectious mercury resistance and its co-transfer with R-plasmid among *E. coli* strains. Indian Journal of Experimental Biology 26: 601-605.

Ajmal, M., Khan, A. (1983). Effects of sugar factory effluent on soil and crop plants. Environmental Pollution 30: 135-141.

Ajmal, M., Khan, A. (1984). Effect of brewery effluent on agricultural soil and crop plants. Environmental Pollution 33: 341-352.

Ajmal, M., Khan, A. (1985). Effect of textile factory effluent on soil and crop plants. Environmental Pollution 37: 131-148.

Alexander, M. (1984). Introduction to soil microbiology. John Wiley & Sons, New York.

Ansari, S.A., Hayat, S., Ahmad, I. (2001). Emergence of high level of metal resistance in the strains of *Rhizobium*. Journal of Microbial World 3: 49-54.

Aziz, O., Inam, A., Samiullah (1999). Utilization of petrochemical industry wastewater for agriculture. Water Air and Soil Pollution 115: 321-335.

Aziz, O., Inam, A., Samiullah, Siddiqi, R.H. (1996). Longterm effects of irrigation with petrochemical industry wastewater. Journal of Environmental Science & Health 31: 2595-2620.

Aziz, O., Inam, A., Siddiqi, R.H. (1994). Impact of treated oil refinery effluent on crop productivity and agricultural soils. Indian Journal of Environmental Health. 36: 91-98.

Baath, E. (1989). Effects of heavy metals in soil on microbial processes and populations. Water, Air and Soil Pollution 47: 335-379.

Bole, J.B., Bell, R.G. (1978). Land application of municipal sewage wastewater: yield and chemical composition of forage crops. Journal of Environmental Quality 7: 222-226.

Brookes, P.C., McGrath, S.P. (1984). Effects of metal toxicity on the size of the soil microbial biomass. Journal of Soil Science 35: 341-346.

Day, A.D., Mc Fadyen, J.A., Tucker, T.C. and Cluff, C.B. (1979). Commercial production of wheat grains irrigated with municipal wastewater and pump water. Journal of Environmental Quality 8: 403-406.

Devarajan, L., Rajannan, G., Ramanathan, G., Oblisami, G. (1994). Effect of one time application of treated distillery effluent on soil fertility status, yield and quality of sugarcane. SISSTA Sugar Journal 20: 133-135.

FAO (1985). Water quality for agriculture. R.S. Ayers and D.W. Westcot. Irrigation and Drainage. Paper 29 Rev. 1. p. 174, FAO, Rome.

FAO (1994). Wastewater treatment and use in agriculture M.B. Pescod, Irrigation and Drainage paper 47. (Ist Reprint in India) Scientific Publisher, Jodhpur, India. ISBN : 81 7233-094-4.

Ghosh, A.B., Bajaj, J.C., Hasan, R., Singh, D. (1983). Soil and water testing methods, A laboratory mannual, IARI, New Delhi, India.

Gupta, I.C. (1988).Quality of irrigation water-Recent criteria and classification. Alfa Publishers and Distributors, Bikaner, India, pp. 1-41.

Hayat, S., Ahmad, I., Azam, Z.M., Ahmad, A., Inam, A., Samiullah (2000). Impact of treated wastewater from an oil refinery on growth and yield of *Brassica juncea* and on heavy metal accumulation in the seeds and soil. Journal of Environmental Studies and Policy 3: 51-59.

Hayat, S., Ahmad, I., Azam, Z.M., Ahmad, A., Inam, A., Samiullah (2002a). Longterm effect of oil refinery wastewater on crops yield, heavy metals accumulation in soil and crop produce. Pollution Research 21: 297-303.

Hayat, S., Ahmad, I., Azam, Z.M., Ahmad, A., Inam, A., Samiullah (2002b). Effect of longterm application of oil refinery wastewater on soil health with special reference to microbiological characteristics. Bioresource Technology 84: 159-163.

Henriette, C., Petitodemange, E., Raval, G., Gay, R. (1991). Mercury reductase activity in the adaptation to cationic mercury phenyl mercuric acetate and multiple antibiotics of gram –ve population isolated from an aerobic fixed bed reactor. Journal of Applied Bacteriology 71: 439-444.

Hughes, M.N., Poole, R.K. (1989). Metal mimicry and metal limitation in studies of metal microbe interaction. In : Poole, R.K. and Gadd, G.M. (Eds.), Metal-microbe Interactions, vol. 26. IRL Press, New York, pp. 1-18.

Inam, A., Saeed, S., Hayat, S. and Ahmad, I. (2002). Effect of wastewater on the heavy metal content of the soil and the seeds of mustard-A case study. In : Saxena, A.K. (Ed.), Heavy metals in the Environment pp. 137-146, Pointer Publisher, Jaipur.

Juste, C., Mench, M. (1992). Longterm application of sewage sludge and its effects on metal uptake by crops. In : Biogeochemistry of trace metals. Advances in Trace Substances Research. Pp. 159-193.

Kansal, F.L., Parwana, H.K., Verma, S.P. (1993). Effect of wastewater irrigation on soil properties. Indian Journal of Environmental Protection. 13: 374-378.

Krishna Murti, C.R., Vishwanathan, P. (1991). Toxic metals in the Indian environment. Tata McGraw Hill Publishing Company Ltd., New Delhi.

Marschner, H. (2003). Mineral nutrition of higher plants. ISBN 0-12-473543-6, Academic press, London.

McGrath, S.P., Brookes, P.C., Giller, K.E. (1988). Effects of potentially toxic metals in soil derived from past applications of sewage sludge on nitrogen fixation by *Trifolium repens* L. Soil Biology and Biochemistry 20, 415-424.

Olson, B.H., Thornton, I. (1982). The resistance patterns to metals of bacterial population of contaminated land. Journal of Soil Science 23: 271-277.

Rajannan, G., Oblisami, G. (1979). Effects of paper factory effluents on soil and crop plants. Indian Journal of Environmental Health 21: 120-130.

Rao, A.V., Jain, B.L., Gupta, I.C. (1993). Impact of textile industrial effluents on agricultural land – A case study. Indian Journal of Environmental Health 35: 132-138.

Samuel, G., (1986). The use of alcohol distillery waste as a fertilizer. Proceeding of International American Sugarcane Seminar, pp. 245-252.

Sandin, G.W., Blender, C.L. (1993). Ecological and genetic analysis of copper and strepto-mycin resistance in *Pseudomonas syringae* pv. Syringal. Applied and Environmental Microbiology 59, 1018-1024.

Smith, S.R. (1991). Effects of sewage sludge application on soil microbial processes and soil fertility. Advances in Soil Science 16: 191-212.

Smith, S.R. (1996). Agricultural Recycling of sewage sludge and the environment. ISBN-085198 9802, CAB International, Wallingford.

Subba Rao, N.S. (1977). Soil microorganisms and plant growth, Oxford & IBH Publishing, New Delhi.

Tyler, G. (1981). Heavy metals in soil biology and biochemistry. In Paull, E.A. and Ladd, J.N. (eds.), Soil Biochemistry, volume 5, Marcel Dekker, Inc. New York, pp. 371-414.

Vanloon, J.C., Lichwa, J. (1973). A study of the atomic absorption determination of some important heavy metals in fertilizers and domestic sewage plant sludges. Environ-mental Letters. 4: 1-8.

CHAPTER 5

Heavy Metals in Temperate Forest Soils: Speciation, Mobility and Risk Assessment

Galina Koptsik[1], Steve Lofts[2], Elizaveta Karavanova[1], Natalia Naumova[3] and Michiel Rutgers[4]

[1]*Soil Science Faculty, Moscow State University, Moscow 119992, RUSSIA,*
[2]*Centre for Ecology and Hydrology, Lancaster Environment Centre, UNITED KINGDOM*
[3]*Institute of Soil Science and Agrochemistry, Russian Academy of Sciences, Siberian Branch, RUSSIA*
[4]*Laboratory of Ecotoxicology, National Institute for Public Health and the Environment, THE NETHERLANDS*

This chapter reviews the available literature about the mobility and bioavailability of heavy metals in contaminated soils in view of environmental risk assessment. The effects of heavy metal contamination on the functioning of soil microorganisms are discussed. The study is focused on heavy metal levels and microbial characteristics in forest soils (Kola Peninsula, Russia), which are heavily polluted by aerial emissions from the nickel-processing industry for several decades. Pathways of heavy metal transformation in soils, the speciation and the distribution of metals between the soil solids and solution are analysed. The most significant factors controlling metal concentration in soil solutions are organic matter (both solid phase and dissolved) and porewater pH. The composition and functioning of soil microbial communities in podzols along the gradient of Ni and Cu contamination is studied by CLPP and other techniques. The conclusion is that the forest soils in boreal region are rather sensitive to metal contamination as adverse effects can be observed already at comparatively low concentrations of contaminants.

INTRODUCTION

Pollution of soil by toxic heavy metals, for example by deposition of industrial emissions or land application of sewage sludge, can cause great harm to terrestrial ecosystems. Effective policy to manage and reduce potentially harmful levels of metals to such systems requires a thorough scientific knowledge of the behaviour and impacts of heavy metals (HM).

The most significant uptake route of metals to soil biota is via the soil solution, and so both the distribution of metal between the soil solids and solution, and the form (speciation) of the metal in the soil solution, are important.

The assessment of mobility and bioavailability is one of the fundamental problems for soils contaminated by HM. It is now generally accepted that for soil organisms exposed via the soil solution, knowledge of the concentration of the free metal ion (FMI) is key to understanding differences in bioavailability and toxicity among soils (Sauvé et al., 1998b; Lofts et al., 2004). HMs may enter the soil in a number of chemical states, for example dissolved in rainwater, complexed to organic matter in sewage sludge, or in mineral forms (for example as sulphide or oxide particles emitted by metal processing plants). Pathways of HM transformation within soils depend on numerous factors, such as inherent soil and environment properties, climate, ground water level, and land use. In the soil, HM may be affected by processes of sorption/desorption, precipitation/dissolution, redox reactions, complexation in porewater, and incorporation into components of the soil matrix or into living tissue, particularly of plants. One distinctive feature of heavy metal contamination of soils is the long residence times of metals, particularly those which bind strongly to the solid phase, e.g. Cu, Pb. This can lead to effects being observed for long periods after removal of the pollution source.

Despite the complex nature of the possible reactions, important factors controlling the distribution of HM between soil and solutes may usually be identified. In temperate forest ecosystems typical of Europe, organic matter (both solid and dissolved (water soluble) and pH are the most significant factors controlling HM solubility and bioavailability.

The environmental risk of HM pollution is pronounced in soils adjacent to large industrial complexes. The Kola Peninsula region, Russia, is heavily impacted by aerial emissions from the nickel-processing industry, specifically the Pechenganikel and Severonikel smelters which are the largest point source emitters of sulphur and heavy metals (particularly nickel and copper) in Northern Europe. Emissions over the last few decades have caused severe damage to the boreal forests of the area, which are among the northernmost coniferous forests of the world and hence may be especially sensitive to impacts of pollution. While the negative effects of air pollutants on forest ecosystems have been studied (Lukina and Nikonov, 1996, 2001; Kashulina et al., 1997; Koptsik et al., 1998, 1999, 2003; Nikonov et al., 2001a, b), no detailed investigations of HM partitioning and the links between soil metals and microbial characteristics have been previously done.

The aim of this study was to investigate the distribution of HM (Ni, Cu, Cd, Pb and Zn) in soils along a pollution gradient in the Kola Peninsula, and the distribution of Ni and Cu between the soil solids and solution. As the studied area has been subject both to HM pollution and sulphur deposition over 50 years, the first goal was to study the effect of pH on metal sorption. Secondly, the impact of dissolved organic matter (DOM) upon the mobility of HM was investigated. In addition, changes in quantitative and qualitative soil bacterial community characteristics were assessed in ecosystems along the pollution gradient.

MOBILITY AND BIOAVAILABILITY OF HEAVY METALS IN SOILS (LITERATURE REVIEW)

Effect of Organic Matter and pH on HM Mobility

Quality criteria for heavy metals in contaminated soils are often based on the total metal in the soil; however, it is clear that for many soil organisms uptake occurs via the soil solution. Given the complex relationship between total metal and dissolved forms in soils, criteria quoted on the basis of total soil metal are not likely to provide good indicators of risk. It has been shown (e.g. Spurgeon and Hopkin, 1996; Lock and Janssen, 2001) that metal toxicity, expressed as total soil metal concentration, varies among soils.

The partitioning of a metal between solid and solution phases of soil is also important for risk assessment because it influences the rate at which the metal is removed from the soil in drainage water. Knowledge of the factors influencing the solid-solution partitioning of metals in soils is therefore, imperative in the study and risk assessment of contaminated land.

The solid-solution partitioning of HMs in soils is governed primarily by complexation, ion exchange and precipitation-dissolution processes. **Complexation** refers to the binding of an ion to a ligand by formation of a specific chemical bond; it includes complexation to the solid phase of soils and complexation with dissolved ligands. Complexation reactions can be expressed by mole balance equations, e.g. $xM + yL = M_xL_y$, where M is the metal ion and L is the ligand (for simplicity, ion charges are not shown). The stability constant, K_c, for such a reaction is given by

$$K_c = \frac{[M_xL_y]}{[M]^x[L]^y}$$

Ion exchange (also termed non-specific sorption) refers to the process whereby solution ions physically accumulate adjacent to a charged surface or molecule, but where specific bonds are not formed. Ion exchange occurs on fixed charge minerals (e.g. clays) but also on variable charge materials such as metal oxides and humic substances. Ion exchange can be described by mole balance equations, e.g. $M1 + [1/n]M2-X_n = M1-X + [1/n]M2$, where M1 and M2 are the exchanging cations (Appelo and Postma, 1993). The integer term n accounts for the exchange behaviour of cations of different valence; for example, if M1 were Na^+ and M2 were Ca^{2+} then $n = 2$. X represents an exchanger charge of -1. The equilbrium constant for such an exchange may be written

$$K_{exch} = \frac{[M1 - X][M2]^{1/n}}{[M2 - X_n]^{1/n}[M1]}$$

Non-specific sorption on humic substances has been modelled (e.g. Tipping, 1994; Benedetti et al., 1996) using variations on the Donnan model.

Precipitation-dissolution processes can be expressed as $M_xA_y = xM + yA$, where M_xA_y is the solid phase formed from the metal M and anion A (ignoring ionic charge). Since the activity in the solid phase is defined as unity, the equilibrium equation is $K_{so} = [M]^x[A]^y$, where K_{so} is the solubility product. This expression defines the concentrations of dissolved ions under circumstances where these are controlled by a solid phase. Precipitation is relevant for many ions but the minerals formed are typically carbonates, phosphates, sulphides and hydroxides that also take part in acid-base reactions.

The overall partitioning of a metal between the solid and solution phases of a soil is governed by a complex combination of the above processes. In addition, concentrations of competing ions (e.g. Ca, Al), ionic strength and the partitioning of natural organic matter, all play a role in determining the overall partitioning.

The Role of Soil Organic Matter in HM Transport in Soils

Soil organic matter (SOM) plays a key role in the transport and fate of HM in soil and groundwater, e.g. as a medium for sorption, a source of ligands, and as a co-substrate for microbial processes. SOM may include low molecular weight acids (aliphatic and aromatic), amino acids, carbohydrates, phenolic compounds, fulvic and humic acids (van Hees and Lundstrom, 1998; Gallet and Keller, 1999; Strobel, 2001). All these substances can dissolve into the soil solution and readily transport HMs to lower horizons or to surface waters (Malcolm and McCracen, 1968). Many authors (Schnitzer, and Scinner, 1967; Bloom, 1981; Stevenson and Welch, 1979; Stevenson, 1994; Johnson et al., 1995, 1997) have proposed that metal transport in forest soils can be largely attributed to co-leaching with DOM. The fate of HM in forest soils may be further influenced by interactions with chemically active phases of underlying mineral horizons, particularly metal oxides.

Of particular note among the constituents of SOM are the humic substances (HS), which are relatively recalcitrant, chemically heterogeneous macromolecules formed as endpoints of microbial breakdown processes and typically form the dominant fraction of DOM in soils. They contain large numbers of functional groups which interact with protons and other soil cations and are consequently considered to be the most important DOM fraction with respect to HM transport and fate.

Concentrations of DOM in Solutions of Temperate Forest Soils

DOM in forest soil solutions derives from two main sources:

1. Directly from throughfall or exudation by plant roots;
2. Decomposition of intact litter input at the soil surface (e.g. needle-fall), to ultimately produce a pool of soil organic matter which can solubilise.

Concentrations of DOM in forest floor horizons are therefore, dependent upon many biological factors, particularly temperature, soil moisture, and the rate and nature of litter inputs. DOM in lysimeter waters of temperate forest floor horizons commonly exhibits large short-term variability, typically with a maximum in late summer (Lundström, 1993; Lukina and Nikonov, 1996; Michalzik and Matzner, 1999; Kaiser et al., 2001). Chemical factors, such as soil pH and the concentrations of major ions, influence DOM by controlling the solubility of the potentially soluble organic matter. For example, Tipping and Woof (1990) studied the effect of pH on the desorption of organic matter from organic soils, finding that pH promoted the solubilisation of organic matter and hence increased DOM. de Wit et al. (2001) studied the solubility of organic matter in a forest floor horizon by manipulating its pH and Al content. For a given Al content, DOM increased at both low and high pH. At high pH, DOM decreased with increasing Al, while at low pH it increased as Al increased. These observations may be partly interpreted by considering the effects of pH and cation concentration upon the hydration of humic molecules. The tendency of a humic molecule to pass from a precipitated to a dissolved form depends on the extent to which it may hydrate, which itself depends on the molecular charge. In a non specifically-binding electrolyte (e.g. NaCl) the molecular charge is determined by the degree of deprotonation, which is pH dependent. Deprotonation gives rise to a negative charge on the humic molecule; mutual repulsion of the charged functional groups then causes the molecule to assume a stretched (hydrophilic) configuration. Binding of protons or other cations reduces the repulsion in the polymer chain which then tends to coil, expelling hydrated water [the Fuoss effect (Ong and Bisque, 1968)]. The molecule becomes more hydrophobic and coagulation or flocculation may occur. This theory does not however explain the observation of de Wit et al. (2001) that DOM increased at low pH in the presence of increased soil Al. Here, it is possible that the initial addition of Al to the soil resulted in the creation of ternary complexes where organic matter was sorbed to the bulk soil solids by a bridging Al ion. Lowering of the pH then caused the

Al-organic matter complexes to desorb from the bulk solids, increasing the DOM.

Other cations besides Al may influence the DOM concentration. Oste et al. (2002) investigated the influence of pH and Ca on organic matter solubility in a number of soils. DOM increased with increasing pH, while Ca addition had the opposite effect. Fotovat and Naidu (1998) studied the effect of major cations and ionic strength (I) on heavy metals (Cd, Cu and Zn) in alkaline sodic and acidic soils. Both I and the nature of the added cation had a marked effect on the concentrations of DOC, Cd, Cu and Zn in solution. With increasing I, DOC decreased in all soils. At a given pH, divalent cations lowered DOC more than did monovalent cations. Associated changes in dissolved heavy metals were detected. While the concentration of Cd increased with I in all soil types, that of Zn increased in acid soils and decreased markedly in alkaline sodic soils. The concentration of Cu also decreased in sodic soils with increasing I. **MINTEQA2** speciation studies showed that the ratio of Zn^{2+} to total Zn increased from less than 1% and 18% to 5% and 76%, respectively, with I in the sodic soils, while the ratio of organic Zn to total Zn decreased from 99% and 78% to 92% and 15% under the same conditions, respectively. This shows that DOM is important in controlling Zn and Cu solubility in these soils, via complexation. Temminghoff et al. (1998) investigated the coagulation of soil humic acid (HA) in response to increasing concentrations of Na, Al, Ca and Cu. At pH 4, coagulation occurred at ~10^{-4} M Al, ~$10^{-3.3}$ M Cu, $10^{-2.5}$ M Ca and above 10^{-2} M Na. Although both Ca^{2+} and Cu^{2+} are divalent cations, coagulation in the presence of Cu occurred at a lower concentration than for Ca, probably because of the stronger binding of Cu by HA. The minimal effect of Na on coagulation is expected since this ion does not specifically bind to HA in significant amounts.

Organic matter in mineral horizons of forest soils is largely derived by leaching from the forest floor above. In contrast to the forest floor, mineral horizons are a net sink of organic matter. Sorption of DOM (and associated cations including HM) may occur by a number of processes. In addition to aggregation/precipitation of DOM by the coagulation process described above for forest floor horizons, many authors believe that direct adsorption of DOM to the charged surfaces of mineral oxides (particularly Al and Fe), as well as sorption at clay surfaces, may play a significant role in OM retention. The relative significance of adsorption and precipitation processes in regulating DOM in mineral horizons is not well understood at present. It is well established that natural organic matter will adsorb to 'clean' mineral surfaces in the laboratory (Mazet et al., 1990; Edwards et al., 1996; Kaiser and Zech, 1997) by surface complexation. It has also been shown that the strength of OM sorption to mineral soils is related to their iron (III) and aluminium oxide content (Kaiser et al., 1996), supporting the theory that sorption to surface sites on minerals controls OM sorption. On the other hand, Jardine *et al.* (1989) proposed aggregation and/or precipitation as a mechanism for OM sorption to soils, based on measured heats of adsorption. Recently Zysset and Berggren (2001) showed that under unsaturated flow conditions (experiments in columns) an illuvial horizon appeared to have no fixed capacity to sorb OM, suggesting that in this case at least aggregation/precipitation mechanisms were important.

The Significance of DOM in Forest Soil Solution Chemistry and Metal Leaching

According to Dijkstra *et al.* (2001) organic acids significantly influenced solution chemistry in the forest floor under six common tree species in the Great Mountain Forest (Connecticut). Between 25 and 43% of the negative charge in solution could be attributed to organic anions. Organic acidity (which correlated with DOC concentration) was balanced by exchange with cations from the cation exchange complex (CEC) and by mineral weathering. The neutralized anions, in return, enhanced leaching of cations from surface soil.

Pohlman and McColl (1988) studied the effect of soluble organics from the forest litter (O horizon) beneath seven tree species, on the dissolution of metals (Al, Fe, Mg, and Mn) from two forest soils. Forest litter was leached by solution with pH 3 in the absence and presence of soil. Soluble organics were split, after Leenheer (1981), into hydrophylic and hydrophobic fractions. HPLC was used for the identification of individual organic compounds. In absence of soil, different litters varied widely in their ability to release dissolved organic carbon (DOC) and metals. No significant correlations between protons consumed and DOC, Al, Mg, and Mn released from forest litter were found. In the presence of soils there was a direct relationship between organics released from litters and Al and Fe mobilized from the soil. The authors suggested that such a correlation confirmed that dissolution of Fe and Al from soil is the result of chelation with certain active components of DOC. A good correlation was obtained between soil dissolved Al and hydrophobic acids, Al and Fe, and hydrophilic acids. Polyphenolic content was highly correlated with dissolved Al and with dissolved Fe. A significant correlation was obtained between total free-organic acid content (oxalic, malic, gallic, protocatechuic acids etc.) and dissolved Al and Fe from soils.

Heavy metal mobilization from polluted soils in the vicinity of a copper smelter was also affected by different organic compounds: citric acid, fulvic acid from high moor peat, and ammonium oxalate (Karczewska, 1999). The study showed that HM desorption from polluted soils may occur in both acidic and alkaline conditions when complexing agents are present. Metal release depended strongly on pH and soil properties. At low pH values metal desorption decreased with increasing pH, reached very low values at slightly acidic pH and usually increased again at higher pH. Addition of soluble organic substances caused certain changes in the desorption results. The most pronounced effects were observed for Fe and Cu solubilization by citric acid in the pH range 3.5 to 6.5, and for Cu solubilization by ammonium oxalate in the pH range 4.5 to 7.

The Mechanisms of DOM Impacts on Metal Mobility and Associated Effects of pH

Organic ligands may indirectly influence metal solubility in two further ways: (i) sorption of the organic ligand to the soil (particularly to metal oxide surfaces), resulting in increased sorption of the metal to the solid phase. This occurs either by complexation of metal to adsorbed organic ligands (ternary complexation), or by enhancement of metal binding to the solid phase as a result of the partial neutralisation of positively charged surfaces, and (ii) competition with the surface for metal complexation, resulting in decreased sorption to the solid phase.

The variable effect of DOM on HM sorption depends on the metal and DOM concentrations, the pH and concentrations of competing cations in the soil solution, the electrical charging properties of the soil, the affinity of the metal for DOM, and the ionic strength. For strongly binding metals, DOM concentration may be a controlling factor in determining their soil solution concentrations.

In the laboratory, the effects of DOM on metal leaching can be studied by manipulation of soils, i.e. by addition of organic ligands under controlled conditions and analysis of leachates. A number of studies have been carried out, some of which are summarised here. An advantage of this approach is that by using simple organic ligands of differing chemical structure (e.g. acetate, oxalate and citrate), some idea of the underlying mechanisms controlling leaching may be elucidated.

Chairidchai and Ritchie (1990) investigated Zn adsorption by a lateritic soil in the presence and absence of different organic ligands (0 to 3 mM of acetate, oxalate, citrate, tricarballylate, salicylate, catechol; 0-3 mM_c of humate). The results of the study showed that metal sorption to the solid phase significantly decreased in the presence of DOC. The authors supposed that pH and competition between soil surface and organic ligands L were the most important factors determining the extent of metal adsorption, which depended largely on the reactions of Zn in solution. In the absence of DOC more than 95% of the Zn was absorbed and the amount of adsorption was linearly correlated with pH and $Zn(OH)^+$ concentration in the equilibrium solutions. Adsorption increased with increasing pH: at

2.5 μmol g^{-1} added Zn, Zn concentration in solution decreased from 238 to 19 μM when pH was adjusted from 3.8 to 5.6. The effect of pH on adsorption in the absence of ligands appeared to affect the forms being adsorbed and/or the mechanism of adsorption, as well as by changing the number of charged sites, although in the given work the amount of Zn was too small to saturate all sites at any of the investigated pH values. In the presence of DOC Zn adsorption and pH decreased, and 74% of the variation in Zn adsorption could be accounted for by the combined effect of $Zn(OH)^+$ and Zn-L concentrations in final solutions. Both hydroxyl ions and organic ligands competed with the soil solids for Zn, so increasing Zn in the soil solution resulted both from the pH change and from metal complexation. The percentage of complexed Zn (as well as Zn adsorption) depended on the affinity of Zn for the organic ligand, the pH, and the relative concentrations of Zn and the ligand. Salicylate and catechol complexed very small proportions of the total Zn. Acetate complexed about 3% to 7% of the Zn, whereas oxalate and citrate complexed almost 100%. For humate the amount of complex formed could not be calculated, but the percentage decrease in Zn adsorption was the largest for all the ligands. At a constant pH, and initial ligands concentration 1 mM the order in which the ligands decreased Zn adsorption was humate > oxalate > citrate > tricarballylate > acetate > salicylate = catechol. It was shown that Zn adsorption could have been affected by formation of negatively charged complexes with oxalate, citrate and tricarballylate. Also the adsorption of ligands may have increased the number of negatively charged sites at the soil surface.

Naidu and Harter (1988) studied the effect of pH and different organic ligands (maleate, citrate, fumarate, succinate, tartrate, malonate, oxalate, salicylate, and acetate) on the sorption and extractability of Cd by four surface soils with widely varying properties. For all soils, the amount of Cd extracted by the organic acid solutions (3 mM) decreased with increasing pH irrespective of the nature of the organic ligand, indicating the importance of pH in controlling Cd availability in these soils. The greatest amount of Cd was released by maleate and the least by acetate. No Cd was released in either nitrate or acetate extracts above pH 6.5. At the same time it was found that even above pH 7, measurable quantities of Cd were detected in ligand extracts. It was suggested that Cd desorption at these pH values is ligand–promoted rather than proton–promoted. Using experimental studies designed to separate the pH effects from ligand ion effects, it was found that at high pH values, Cd–ligand complexation was essential for the solubilization of Cd.

The amounts of Cd in acetate, oxalate, citrate, and nitrate extracting solutions of varying concentrations in pH-buffered solutions were compared. Extractable Cd at a given pH depended on both the concentration and the nature of the ligand ion. Of the ligand ions investigated, changes in the concentration of nitrate had only a minor effect on extractable Cd at a given pH value. The organic ligand solutions, however, gave contrasting trends depending on the solution pH. At low pH, increasing the oxalate or citrate concentration resulted in a decrease in extractable Cd. In contrast, at pH values above 6, increased ligand concentrations increased extractable Cd. Acetate-extractable Cd was not consistently correlated with concentration. Two factors could control Cd adsorption in the presence of ligands. First, at low pH when the oxidic soil surfaces were positively charged, ligand adsorption may have led to charge reversals:

$$S^{m+} + L^{n-} = [S-L]^{m-n};$$

where S^{m+} represents a positively charged surface binding site, the charge of which may be greater than +1 if it is formed from more than one functional group. Such charge reversals (supported by zeta potential measurements) could enhance adsorption of metal ions. Secondly, formation of negatively charged complexes may have enhanced adsorption onto positively charged surfaces by ternary complex formation:

$$M^{p+} + L^{m-} = M-L^{p-m};$$
$$M-L^{p-m} + S^{n+} = [S-(M-L)]^{n-(p-m)}$$

The relative extents to which these two processes enhance the adsorption would depend upon a number of factors, of which the influence of surface charge on metal binding affinity, and the strength of ternary metal-ligand-surface binding are likely to have been most important. The oxalate ion, which possesses two binding functional groups, is in theory able to form ternary complexes with a metal cation and an oxide surface, whereas acetate, which has only one functional group, cannot form ternary complexes, and is thus less efficient than oxalate at decreasing Cd leaching at low pH. The structure, as well as the chemistry, of the organic ligand, is key to its ability to mobilize metal. Structure is especially important when considering humic substances, since these polyfunctional macromolecules may not only form ternary complexes but readily bind metals very strongly by chelation at multidentate binding sites.

Elliot and Denneny (1982) studied adsorption of Cd by soils, using solutions containing organic ligands under typical field pH conditions. Half or more of the soluble Cd was present as Cd^{2+} while the remainder, depending on the system, existed as $CdAc^+$, $CdOx^+$, $CdNTA^-$, or $CdEDTA^{2-}$. Organic complexation either reduced or had an imperceptible effect on Cd adsorption relative to free Cd^{2+}. In general, the ability to reduce sorption followed the order of Cd-ligand stability constants EDTA > NTA > oxalate = acetate. Thus, the strength of the Cd-ligand complex could have been one of several parameters controlling adsorption. As the strength of the Cd-ligand bond increased, the ability of the organic ligand to compete for Cd also increased. Reduction in sorption in the presence of citrate was consistent with the fact that citrate forms a strong complex with Cd.

The effect of Cu/DOC ratio and pH on copper speciation and mobility was analysed by Nierop $et\ al.$ (2002). The interaction of Cu with dissolved organic matter (DOM), extracted from an organic forest floor, was investigated. The speciation of Cu over 'free' Cu (as analysed by diffusive gradients in thin films (DGT)), dissolved Cu–DOM complexes and precipitated Cu–DOM was determined as a function of pH (3.5, 4.0 and 4.5) and Cu/C ratio. The dissolved organically bound fraction was highest at pH 4.5, but this fraction decreased with increasing Cu/C ratio, which was observed for all pH levels. In the range of $Cu/C = 7 \times 10^{-5} - 2.3 \times 10^{-2}$ (M/M) the precipitated fraction was very small.

Sauvé $et\ al.$ (1998a) investigated the effects of SOM content and pH on the solubility of Pb in contaminated soils. Total dissolved Pb and free Pb^{2+} were investigated in soils treated by leaf compost and H_2O_2, yielding six soil organic matter levels between 26 and 84 g C kg^{-1}. The range of pH investigated was 3 to 8. The solubility of organic matter increased with higher SOM and pH above 7. For soils with the highest SOM increases in DOC were also observed at lower pH:

$$DOC = 68.2 - 30.8pH + 2.54pH^2 + 14.9SOM,$$

where DOC is in milligrams C per liter and SOM is in grams C per kilogram.

The solubility of Pb increased from pH 6 to pH 3: the relationship between dissolved Pb (pPb) and pH was practically linear. The near constant or increasing Pb solubility at higher pH could be attributed to the increasing proportion of Pb hydroxy- and carbonate species and especially Pb-DOM complexes in solution. At near neutral pH the activity and concentration of dissolved Pb showed no significant effect of pH, and a small but significant effect of OM (higher SOM increases DOM and promotes the formation of Pb-DOM complexes, increasing Pb solubility) and above pH 6 labile Pb showed a practically linear increase with DOC:

$$pPb = 2.18pH - 0.14pH^2 - 0.024DOC - 1.05$$

Dynamic Studies of HM Transport in Soils

A few dynamic studies of HM transport in soils exist. As an example, Temminghoff $et\ al.$ (1998) studied the effect of the properties and state of DOM on the mobility of copper in a contaminated sandy soil. Two soils were used in duplicate for the column experiment. Columns were leached with sodium nitrate (0.003 M) for an initial period, following which one column of each soil was leached with a solution of sodium nitrate (0.003 M) and DOM (purified HA; 185 mg l^{-1}). Both DOC and dissolved Cu increased appreciably in percolates of these columns while pH and Ca remained relatively unaffected compared to the control columns, although there was a slight increase in Ca in percolates of the more alkaline soil. Following this period, the sodium nitrate in all the feed solutions was replaced with calcium nitrate (0.0005 M) adjusting the ionic strength to 0.003 M with sodium nitrate. In the columns with DOM in the feed solution, Cu and DOM in the percolates dropped due to the addition of Ca. In the more acidic soil, Cu and DOM were depressed to levels similar to those seen in the control, while in the more alkaline soil levels remained higher than those in the control. In the final period, the calcium nitrate concentration in the feed of the more acidic soil columns was decreased (0.0001 M) while in the feed of the more alkaline soil columns it was increased (0.0025 M). In the more acidic soil, percolate concentrations of Cu and DOM gradually increased but did not reach the concentrations previously observed, while in the more alkaline soil, percolate Cu and DOM dropped to very low concentrations. The authors concluded that Cu mobility was partly regulated by DOM concentration and partly by the Ca concentration. Three processes controlling Cu leaching were identified: desorption from the soil solids, DOM coagulation, and complexation. These processes were in simultaneous action, and all three were influenced by pH and Ca. At low pH desorption from soil appeared dominant, whereas at high pH complexation by DOC dominated. Cu in the percolates was simulated by the NICA model (Koopal et al., 1994), a sophisticated model of humic substance chemistry incorporating binding site heterogeneity and able to account for competition among H^+, Ca^{2+} and Cu^{2+} for HA. Adjustment of two parameters of the model (maximum Cu adsorption capacity and intrinsic heterogeneity) allowed the prediction of percolate Cu over a range of pH, Ca and Cu concentrations.

The Influence of Heavy Metal Pollution on Soil Bacterial Communities

At higher concentrations HMs are toxic to all living organisms, their most common impact mechanisms being inhibition of enzymes or binding to DNA. Even highly reputable trace elements like Zn^{2+} or Ni^{2+} and especially Cu^{2+} are toxic at higher concentrations. Soil microorganisms, and especially bacteria and archaea, directly interact with organo-mineral surfaces of the soil matrix and need the soil solution to thrive. Hence soil bacteria and archaea are directly exposed to the effects of increased heavy metal concentrations in soil, which may affect bacterial community composition and functioning.

Estimation of the structural and functional diversity of microbial communities in various environmental habitats has been for some time the focus of microbial ecological research, since microbes in both aquatic and terrestrial ecosystems are responsible for fundamental ecosystem processes. For the assessment of functional aspects, community-level physiological profiles (CLPP) generated with sole-carbon-source-utilisation tests using BIOLOG® microplates have become especially popular. These tests can provide physiological data on the soil bacterial population under different environmental and/or management practices.

The effect of HMs on soils depends on the duration and mechanism of HM input into an ecosystem, as well as ecosystem type and land use. It has been established by now that the long-term effects of HMs on soil differ from the short-term effects (Renella *et al.*, 2002). The long-term effects of HMs are of particular concern since their interference in bacteria-mediated processes may affect the overall quality of the biosphere.

Little is known about the changes in microbial diversity associated with ecosystem degradation due to adverse effects. There are few studies devoted to the long-term effects of HMs on soil ecosystems, mostly with sewage sludge addition to agricultural ecosystems (Pennanen *et al.*, 1996; Knight *et al.*, 1997; Bååth *et al.*, 1998). There are few sharp gradients in the world of aerial deposition of HMs, and such deposition is difficult to model in the laboratory or the field, especially on a long-term basis. Little is known about the functioning of soil microorganisms in ecosystems exposed to long-term aerial contamination by HMs.

Conclusions

Soil solution pH and DOC concentrations can have appreciable impacts on the mobility of HMs in forest soils. The effects are however complex and dependent upon many factors including the concentrations of other competing ions, the active chemical phases present on the soil solids, the concentration and structure of the DOM, and the affinity of the HM and competing ions for the soil and solution phases. Laboratory manipulation of soils, for example by addition of organic ligands, may provide valuable information on the mechanisms controlling dissolved metals. For the integration of findings and for dynamic studies of metal mobility, speciation modelling (Temminghoff *et al.*, 1998) has a valuable role to play. However, site-specific or empirical models also have a valuable role where the large amounts of information required to run speciation models are not available. The composition and activity of soil microbial communities in heavy metal contaminated ecosystems may decrease as well, the effect in field studies being partly caused by other than HM factors.

EXPERIMENTAL

Study Area

The study area is located around the Pechenganikel and Severonikel smelters in the Kola Peninsula, north-western Russia (Fig. 5.1). It is a hilly subarctic glaciated terrain covered by tills, fluvioglacial deposits and open bedrock with coarse texture. The glacial till mineralogy of the study area is generally characterized by slowly weathered primary minerals, represented by mostly quartz and feldspars derived from gneisses and granites. The area is located at the polar tree line. Pine and spruce forests are characterised by scarce tree stands and scarcely developed ground vegetation, dominated by crowberry, blueberry, mosses and lichens (see also Lukina and Nikonov, 2001).

Fig. 5.1   Study area.

Podzol is the most common soil type, ususally of less than 50 cm thickness. Podzols are characterized by the presence of two powerful geochemical barriers: organic surface horizons of the forest floor (O), and mineral illuvial horizons (B_{sh}). These horizons differ very much in their properties, including their DOM content. The forest floor plays a very important and dual function in soil HM chemistry. The chemically active organic matter produced by partial forest floor degradation can strongly retain HMs but when dissolved can transport HMs to lower horizons and to surface waters. Overall the high concentration of active organic matter in the forest floor reduces the free metal ion (FMI) activity, which is generally considered a good indicator of bioavailability and toxicity to soil organisms. All these peculiarities have been taken into account in this study, as they are very important in assessing the environmental risk of HM pollution.

Sampling

Soil samples were collected from spruce and pine forests represented background, defoliating, and sparse forest states and from "industrial barren" at different distances from the smelters. Separate soil samples were taken from the organic (O, mostly 2-6 cm) and different mineral (E, B_{sh}, BC and C) horizons. Besides, samples of T horizon of peat gley soil (S-41 plot) and A horizon of agricultural soil (P-60 plot) also were used in the study. For chemical analysis, the soil samples were air dried at 20°C and sieved with a 2 mm sieve.

Chemical Methods

Particle size distributions were determined by the pipette method and sieving. Fe and Al (Fe_o, Al_o) were extracted with oxalate after Vorob'eva (1998). pH in H_2O and 0.01 M $CaCl_2$ suspensions was measured at 1:25 and 1:2.5 soil/solution mixtures for organic and mineral horizons, respectively. Total organic carbon was measured by a dry combustion method using an express analyzer AN-7529. "Total" (Aqua Regia) and available (0.43 M HNO_3 and 1 M NH_4NO_3) metals (Ni, Cu, Cd, Pb, Zn) were determined according to ISO 11466:1995 and Houba et al. (1985). Metals were analyzed by atomic absorption spectroscopy. All element concentrations are presented on a dry matter basis.

Soil sorption of copper and nickel was studied in batch experiments, where experimental conditions were approximated to the field situation: low ionic strength of the soil solutions and relatively low concentrations of Cu and Ni. Soil contamination was simulated in two ways. In the first two experimental runs sorption of the metals was studied in the absence of competition, so copper and nickel were added to the soils singly as water soluble salts [copper sulphate ($CuSO_4$) and nickel sulphate ($NiSO_4$)]. In the third experimental run copper and nickel sulphates were jointly added to the soils in order to investigate competition effects. Experiments were carried out at three different initial pH levels: 5.5 (5), 4.0 and 3.0 adjusted with HCl. The experiments were carried out at constant temperature (20°C), initial ionic strength (0.015 M) [controlled by 0.005 M $Ca(NO_3)_2$] and at a soil-solution ratio of 1:10. Metal concentration in the input solutions (loading) were: 0.01, 0.04, 0.08, 0.2, 0.6, 1.0 mM. Experiments were carried out in triplicate. Suspensions were shaken for 2 hours, than equilibrated for 48 hours. After the equilibration period suspensions were filtered through 0.45 μm membrane filters. Copper and nickel concentrations (by atomic adsorption spectroscopy, AAS), pH and copper activity (by ion–selective electrodes, ISE) were measured in the equilibrium solutions. DOC concentrations were measured with a Hach COD Reactor. Metal concentrations in the equilibrium solutions, copper activities, pH and DOC concentrations were used in multiple regression analysis.

Microbiological Methods

After sampling the forest floor was manually homogenized by cutting into small (ca. 0.5 cm) pieces, mixing, and preincubating at 50% WHC for 4 weeks at 10°C prior to analysis.

Preparation of slides for microscopic counting of bacteria was performed according to the procedure described by Bloem (1995). An aliquot (9 ml) of soil supernatant (see below) was taken into a 10-ml polypropylene tube and vortexed with formalin (1 ml, final concentration of 3.7%). To make slides, suspensions were vortexed again and after 2 min of sedimentation 10 μl of suspension was put onto a glass slide and distributed evenly in a circle of area 113 mm^2. After air-drying, 30 μl of stain solution, containing 0.2 mg $\times ml^{-1}$ of DTAF in the buffer (0.05 M Na_2HPO_4 and 0.85% NaCl, pH 9.0), was added to the slides. The latter were incubated for 30 min in the dark, and then rinsed 3 times for 20 min in the same buffer, followed by rinsing for a few moments in demineralized water. Counting was performed with a confocal laser-scanning microscope (CLSM, Leica Lasertechnik GmbH, Heidelberg, Germany) with an automatic image analysis system (Bloem et al., 1997).

To characterise metabolic profiles of bacterial communities with the CLPP technique (Garland, 1997; Konopka et al., 1998) aliquots of soil were suspended (1:10 w/v) in a sterile 10 mM phosphate buffer (pH 7.0), containing 9 gl^{-1} sodium chloride by blending during 1 min at 20.000 rev min^{-1}. The suspensions obtained were centrifuged for 20 min at 5,000 g to remove fungal mycelia, coarse organic and mineral particles. 12 serial 3-fold dilutions of the obtained supernatant were then prepared. Each dilution (in 0.1 ml aliquots) was used to inoculate 96-well microtiter Biolog® ECOplates, containing triplicates of 31 different organic substrates and of a control without any substrate (Biolog Inc., Hayward CA. USA) as well as a tetrazolium redox dye and a dried mineral salts medium (Preston-Mafham et al., 2002). If bacteria are able to utilise a certain C source, they respire; this results in the reduction of tetrazolium dye in a well and development of violet colouring. Optical density developed in course of incubation at 20ºC and 98% humidity was measured at 590 nm automatically, three times a day for a week with a plate photometer (Garland *et al.*, 2001). Thus curves of the absorbance dynamics for each substrate and each dilution of soil suspension for every soil sample, were obtained. Per inoculum dilution, the colour development curves for each of the substrates as well as the averaged absorption over all substrates (average well colour development, AWCD) were integrated over time to give the area under the curve (AUC). The AUC for each substrate was plotted against log numbers of bacterial cells in a well, a log-logarithmic curve fitted, and the log number of cells which resulted in 50% of the maximal AUC ($EC50_{well}$), calculated for each well as well as for all the wells in average ($EC50_{AWCD}$) by the non-linear estimation module of the Statistica v.5.5 package.

Part of the dilution series was plated onto tryptic soy agar (Oxoid). Plates were incubated at 20°C, and the number of colony-forming units (CFU) was counted daily from day two to day eight. The percentage of fast-growing bacteria was determined by dividing the number of CFU counted on day two by the total number of CFU on day eight.

All the data obtained were log-transformed, and discriminant analysis, principal components analysis, analysis of variance and significance tests for the differences between the sites (Student's t-test) were performed using the corresponding modules of Statistica v.5.5.

RESULTS AND DISCUSSION

Soil Chemistry

The soils in the study area are mainly coarse sandy podzols with low clay contents (3-6% in E, 6-9% in B_{sh} and 3-7% in C horizons, Table 5.1).

The distribution of oxalate-extractable Al and Fe within mineral profiles was eluvial-illuvial. Significant amounts of Al and Fe oxides were accumulated in the B_{sh}-horizons. In polluted soils, Al tended to be concentrated in the lower mineral horizons compared to the background soils. Fe and Al contents in O-horizons increased near the smelter due to technogenic input and erosion, while in B_{sh} and C horizons they depended mainly on lithology.

The content of organic carbon in the O-horizon varied from 32 to 42%, in the B_{sh}-horizon from 0.5 to 4%, and in the C-horizon from 0.1 to 0.6%. The content of organic matter in the topsoil decreased near the smelter due to decreased litterfall input in sparse forests.

Table 5.1   General properties of podzols along the pollution gradient in the Kola Peninsula

Distance	Horizon	Depth	Clay	Al_o	Fe_o	C_{org}	N_{tot}	pH_{H_2O}	pH_{CaCl_2}
km		cm	%	%	%	%	%		
200	O	0–7	–	0.046	0.12	40	1.3	4.1	3.4
	E	7–12	6.0	0.018	0.01	0.9	0.05	4.1	3.4
	B_{sh}	12–18	11.0	1.5	2.7	2.1	0.12	4.6	4.3
	BC	18–24	3.7	0.87	0.30	0.8	0.05	5.1	4.7
	C	24–50	5.1	0.28	0.07	0.3	0.02	5.1	4.6
100	O	0–9	–	0.08	0.07	41	1.4	4.1	3.3
	E	9–14	4.0	0.02	0.02	1.0	0.02	3.6	3.3
	B_{sh1}	14–19	6.7	0.70	0.74	1.5	0.08	4.3	3.9
	B_{sh2}	19–30	4.2	0.71	0.15	0.5	0.04	5.0	4.6
	BC	30–42	4.1	0.88	0.20	0.5	0.01	5.0	4.8
	C	42–50	4.8	0.17	0.03	0.1	0.02	5.2	4.8
28	O	0–7	–	0.07	0.09	36	1.6	3.9	3.2
	E	7–12	4.0	0.02	0.03	0.8	0.05	3.9	3.3
	B_{sh1}	12–19	6.3	1.6	1.9	4.0	0.23	4.4	4.1
	B_{sh2}	19–28	4.0	1.6	0.79	2.0	0.13	4.6	4.5
	BC	28–52	3.4	1.02	0.23	0.6	0.04	5.0	4.8
	C	52–70	3.8	0.33	0.05	0.2	0.03	5.0	4.8
20	O	0–6	–	0.31	0.22	35	0.69	4.2	3.4
	E	6–10	4.9	0.04	0.07	2.7	0.16	4.1	3.3
	B_{sh1}	10–26	7.7	0.59	1.1	2.7	0.14	4.5	4.0
	B_{sh2}	26–33	7.6	1.2	0.95	4.2	0.17	4.8	4.3
	BC	33–41	4.2	1.0	0.20	1.3	0.08	5.8	5.2
	C	41–62	5.0	1.3	0.22	0.8	0.05	5.8	5.2
7	O	0–4	–	0.19	0.36	32	1.1	3.9	3.1
	E	4–6	4.5	0.05	0.06	0.9	0.05	4.1	3.3
	B_{sh}	6–23	5.7	2.1	0.81	4.3	0.21	4.9	4.4
	BC	23–29	3.8	2.2	0.40	1.3	0.11	5.5	5.1
	C	29–60	4.1	0.75	0.06	0.6	0.04	5.6	5.0
2	O	0–3	–	0.47	0.58	26	0.79	4.0	3.3
	E	3–8	5.2	0.05	0.03	2.0	0.11	4.2	3.8
	B_{sh1}	8–28	8.7	0.28	0.24	2.8	0.13	4.5	4.1
	B_{sh2}	28–44	9.7	0.79	0.32	1.5	0.16	4.8	4.4
	BC	44–83	4.8	0.45	0.19	1.1	0.11	5.3	4.9
	C	83–95	9.3	0.37	0.11	0.4	0.01	5.4	4.9

The studied podzols are rather acidic, due to the moist, cool climate, base-poor granite bedrock and mainly coniferous forest cover (Table 5.1). They are characterised by low pH, elevated exchangeable acidity, low contents of exchangeable cations and low effective CEC. The pH usually increases with depth. Despite high sulphur emissions and deposition there were no significant differences in pH values along the pollution gradient. Both direct and indirect influences of air pollution determine changes in pH and exchangeable cations in the study area: 1) increased inputs of base cations from the atmosphere near the pollution sources; 2) possible replacement of H^+ and base cations by pollutant ions (Ni^{2+}, Cu^{2+}, Fe^{3+}); 3) decreased inputs of H^+ due to reduced decomposition of forest floor organic matter and by release of H^+ as a result of reduced base cation uptake by plants in heavy damaged forests and industrial barrens; 4) inherent geological variability (Koptsik et al., 1999).

Heavy metals were significantly concentrated in the surface O horizons (Fig. 5.2 and 5.3, see also Nikonov et al., 2001). Soils close to the smelters contained higher concentrations of nickel, copper, cadmium and lead compared to remote sites. At the most exposed site aluminium, iron, cobalt, chromium, strontium, arsenic and other elements were also found at elevated concentrations (Reimann et al., 1998). Close to the smelter the total concentrations of nickel and copper in the O horizon were in the range 2300-5700 mg kg^{-1} and 2000-3000 mg kg^{-1}, respectively. At remote sites (more than 200 km from the smelter), total concentrations of metals were usually below 10 mg kg^{-1}. Ni and Cu pollution is observed at least 100 km from the Severonikel smelter.

Cd concentrations in the O horizon were over an order of magnitude higher that those observed in B_{sh}- and C-horizons. The O horizon also had the highest levels of Pb. While concentrations of nickel and copper in O horizons were two to three orders of magnitude higher than in mineral horizons, the differences were smaller for Cd and Pb. Concentrations of Zn decreased towards the smelter (data not shown) probably due to replacement by the main polluting metals, and to leaching from the topsoil.

The determination of the different metals in solution is important to evaluate their behaviour in the environment and their mobilisation capacity. The environmental impact of heavy metals depends mainly on mobile, bioavalable forms. Extraction with 0.43 M HNO_3 isolates the pool of heavy metals from soil which controls the soil solution concentration and free ion activity. Acid-soluble Ni and Cu accounted for 30 and 70% of the "total" content in the organic horizons and 5-30% and 10-80%, respectively, in mineral horizons (Table 5.2).

Ammonium nitrate extraction was used to remove from the soil metals predominantly in exchangeable forms. Ammonium nitrate extractable Ni and Cu accounted for 3% to 9% and up to 7% of the totals in the organic and mineral horizons, respectively.

The mobility of Ni and Cu in the soils was limited, as shown by a sharp reduction in their concentrations with soil depth (Fig. 5.2 and 5.3). However,

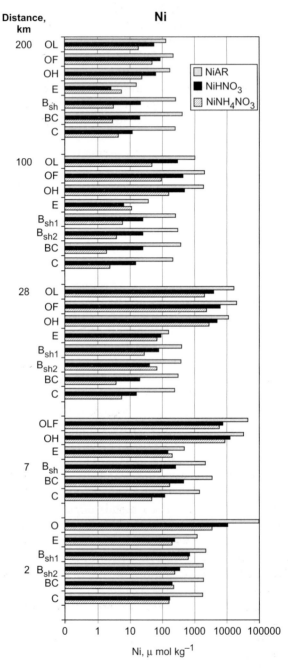

Fig. 5.2   Contd.

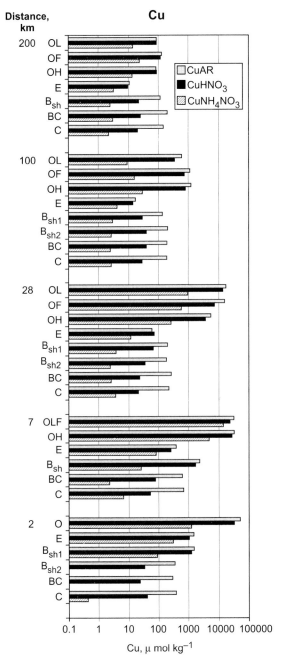

Fig. 5.2 Ni and Cu concentrations (log-scale) in podzols along the pollution gradient.

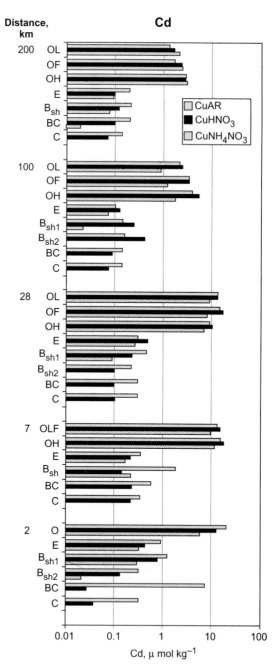

Fig. 5.3 Contd.

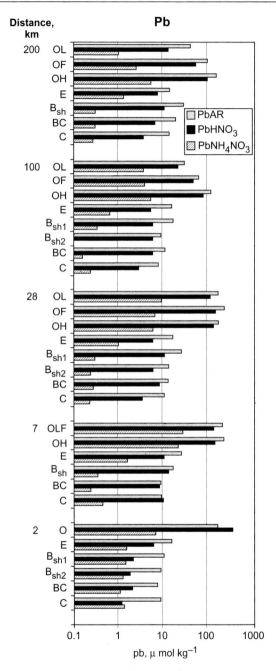

Fig. 5.3 Cd and Pb concentrations (log-scale) in podzols along the pollution gradient.

Table 5.2   Mobile versus total heavy metals in podzols (median values)

Horizon	Ni	Cu	Zn	Cd	Pb
		0.43 M HNO$_3$ vs AR			
O	29	74	79	104	73
E	34	82	14	78	38
B$_{sh}$	14	32	7	55	31
BC	5	12	5	34	25
C	5	11	3	30	25
		1 M NH$_4$NO$_3$ vs AR			
O	9	3	28	42	4
E	29	17	10	65	6
B$_{sh}$	7	1	3	17	8
BC	1.2	0.03	0.9	4	11
C	1.2	0.3	0.9	3	8

the concentrations of Ni and Cu in mineral horizons were highest in sites in the vicinity of the smelter. In contrast to the organic horizons, the fractions of acid-soluble Ni and Cu and ammonium nitrate extractable Ni in the mineral horizons increased near the smelter.

Organic Matter and pH as Factors in Cu and Ni Mobility in Podzols

General characteristics of soils in the adsorption study are shown in Table 5.3. The O and B$_{sh}$ horizons are characterized by different sorption properties and parameters (Table 5.4, see also Vologdina *et al.*, 2004). Forest floors (O horizons) show the greater sorption affinities for both metals. Under the experimental conditions complete saturation of the O horizon materials by copper was not observed: the isotherms were approximated in most cases by linear functions (Fig. 5.4) described by the Henry equation. The isotherms of Ni sorption varied in form and in most cases they were curvilinear (Fig. 5.5) indicating a lower sorption capacity for Ni. In the low concentration range, the amounts of Cu sorbed by O and B$_{sh}$ horizons did not differ greatly. Such differences became evident at the highest input metal concentrations. Where Cu was added singly, the difference in Q$_{Cu}$ between horizons (Table 5.5) ranged from 12 to 20%. With competition these differences increased to 48%.

Nickel sorption by the O horizons exceeded the sorption potential of the B$_{sh}$ horizons at all concentrations. In the absence of competitions the Ni sorbed by O horizons was 43-47% greater than that adsorbed by the B$_{sh}$ horizons (K-200, Table 5.5). Under competing conditions Ni sorption by the O horizons was 40-80% greater than that of the B$_{sh}$ horizons.

Table 5.3 Properties of soils used in the batch adsorption experiments

Plot	Vegetation	Parent material	Soil	Horizon	C_{org} %	Clay %	pH_{H_2O}	pH_{CaCl_2}	Ni_{HNO_3}	Cu_{HNO_3}
									\u03bcmol/kg	
K–200	spruce forest	unsorted till	podzol	O	40	–	4.0	3.5	70	100
				B_{sh}	2	11	4.5	4.1	20	22
S–41	pine forest	sorted till	podzol	O	36	–	3.9	3.2	770	300
				B_{sh}	2	2	4.5	4.1	35	30
S–16	pine forest	unsorted till	podzol	O	35	–	4.0	3.3	3900	2630
				B_{sh}	3	4	4.5	4.0	80	30
S–41	peat bog	sorted till	peaty-gley high moor soil	T	42	–	3.9	3.3	130	60
P–60	grass	sorted till	agricultural soil	A	9	0.5	6.1	5.4	50	90

Table 5.4   The parameters of Cu and Ni sorption (K-200) approximated by Langmuir, Freundlich and Henry equations

Horizon	Version	Langmuir		Freundlich		Henry
		Q_{max}	k	S_o	$1/n$	K_H
		Cu				
O	$CuSO_4$ pH=5.5	–	–	–	–	264.5
B_{sh}		10.7	10.3	17.1	0.54	–
O	$CuSO_4+NiSO_4$, pH 5.5	–	–	–	–	258.4
B_{sh}		12.0	11.8	23.5	0.59	–
O	$CuSO_4+NiSO_4$, pH 4.5	–	–	–	–	222.2
B_{sh}		12.1	14.8	26.0	0.58	–
O	$CuSO_4+NiSO_4$, pH 3.5	–	–	–	–	203.5
B_{sh}		15.5	4.3	21.6	0.69	–
		Ni				
O	$NiSO_4$, pH=5.5	32.9	1.7	35.2	0.87	44.2
B_{sh}		23.1	0.4	7.4	0.94	7.8
O	$NiSO_4$, pH=4	25.7	2.4	34.1	0.85	45.4
B_{sh}		16.9	0.6	7.5	0.89	8.2
O	$NiSO_4$, pH=3	25.8	2.6	34.8	0.84	47.0
B_{sh}		18.9	0.5	7.2	0.92	7.7
O	$CuSO_4+NiSO_4$, pH 5.5	21.8	2.3	25.1	0.81	34.0
B_{sh}		22.9	0.3	4.9	0.92	5.1
O	$CuSO_4+NiSO_4$, pH 4.5	18.3	3.1	24.1	0.77	35.4
B_{sh}		11.4	0.7	5.3	0.84	5.8
O	$CuSO_4+NiSO_4$, pH 3.5	15.0	3.6	19.7	0.73	29.8
B_{sh}		–	–	5.6	1.00	5.6

The soils also exhibited different "affinities" for Ni and Cu. Values of the selectivity coefficient $D_{Cu-Ni}= \dfrac{K_{H,Cu}}{K_{H,Ni}}$ varied in the organic horizons from 6.3 to 24.3. Here $K_{H,Cu}$ and $K_{H,Ni}$ represent Henry constants for Cu and Ni, respectively.

The B_{sh} horizons also showed different adsorption characteristics for Ni and Cu. Copper sorption was described only by Langmuir (Freundlich) equations (Fig. 5.4) whereas some Ni sorption isotherms had a linear form, depending on the experimental conditions and the soil (Fig. 5.5). Example isotherm parameters for K-200 are shown (Table 5.4). The contrast between Cu and Ni sorption strength was greater in the B_{sh} horizons than in the O horizons. The amounts of sorbed Ni (Q_{Ni}) differed from the amount of the sorbed Cu (Q_{Cu}) by 46-18% in the B_{sh} horizons. In the O horizons Q_{Ni} differed from Q_{Cu} by 53-83%.

Table 5.5   Copper and nickel sorption by podzols of the Kola Peninsula

Plot	Horizon	Variant	$K_{H, Cu}$	$K_{H, Ni}$	$D=K_{H, Cu}/K_{H, Ni}$	$Q_{max, Cu}$	$Q_{max, Ni}$	Sorbed metals at $C_{in}=0.6$ mM		Sorbed metals at $C_{in}=1$ mM	
								Q_{Cu}	Q_{Ni}	Q_{Cu}	Q_{Ni}
K–200	O	CuSO₄	264.5	–		not calc	–	5.79	–	9.63	–
	B_sh	pH 5	not calc	–		10.7	–	5.05	–	7.66	–
	O	NiSO4	–	44.2			32.9	–	4.96	–	8.11
	B_sh	pH 5	–	7.8			23.1	–	2.83	–	4.28
	O	CuSO₄+NiSO₄	258.4	34.0	7.6	not calc	21.8	5.79	4.77	9.62	7.63
	B_sh	pH 5	not calc	5.1		11.9	22.9	5.26	2.06	8.27	3.34
	O	CuSO₄+NiSO₄	222.2	35.4	6.3	not calc	18.3	5.74	4.77	9.57	7.7
	B_sh	pH 4	not calc	5.8		12.1	11.4	5.44	2.33	8.45	3.6
	O	CuSO₄+NiSO₄	203.5	29.8	6.8	not calc	15.0	5.73	4.78	9.52	7.35
	B_sh	pH 3	not calc	5.6		15.5	–	4.83	2.12	7.77	3.59
S–41	O	CuSO₄ pH 5	343.8	–		not calc	–	5.84	–	9.72	–
	B_sh		not calc	–		8.03	–	3.93	–	5.84	–
	O	CuSO₄+NiSO₄	350	17.9	19.6	not calc	15.2	5.85	4.19	9.72	6.25
	B_sh	pH 5	not calc	2.2		8.4	30.3	4.37	1.11	5.74	1.82
	O	CuSO₄+NiSO₄	299	12.3		not calc	14.1	5.96	3.75	9.65	5.3
	B_sh	pH 4	not calc	2.9	24.3	7.4	8.0	3.85	1.46	5.61	2.15
	O	CuSO₄+NiSO₄	296.7	–		not calc	No data	5.84	–	9.66	–
	B_sh	pH 3	not calc	1.2		6.4	2.4	3.11	0.75	4.97	0.99
S–16	O	CuSO₄ pH 5	188.5	–		not calc	–	5.69	–	9.58	–
	B_sh		not calc	–		10.9	–	4.98	–	7.7	–
	O	CuSO₄+NiSO₄	159.2	10.8	14.7	not calc	6.3	5.7	3.35	9.4	5.51
	B_sh	pH 5	not calc	2.7		11.4		5.28	1.35	8.2	2.08
	O	CuSO₄+NiSO₄	155.8	10.8	14.4	not calc	7.8	5.68	3.38	9.4	5.48
	B_sh	pH 4	not calc	2.8		11.3		5.27	1.38	8.17	2.09
	O	CuSO₄+NiSO₄	142.8	8.1	17.6	not calc	16.5	5.67	3.01	9.34	4.94
	B_sh	pH 3	not calc	2.3		12.7		4.61	1.15	7.36	1.82

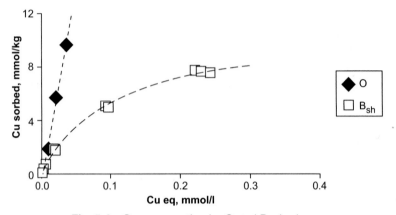

Fig. 5.4 Copper sorption by O and B$_{sh}$ horizons

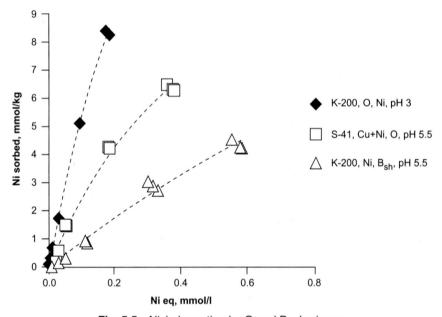

Fig. 5.5 Nickel sorption by O and B$_{sh}$ horizons

The two types of horizon also showed significant variations in the concentrations of free copper ions (see also Vologdina *et al.*, 2004). This may be attributed to the higher soil organic matter in the O horizons, which complexes Cu and hence reduces the free ion activity, while at the same time increasing the total dissolved Cu by releasing DOM to the solution. Copper activity in the forest floor solutions was one to two orders less than the B$_{sh}$ horizons (Fig. 5.6), while total dissolved Cu concentrations differed only by

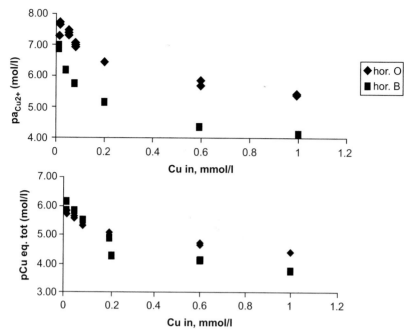

Fig. 5.6 Copper equilibrium concentration and activity in soil solutions (plot K-200, Cu+Ni, pH in=5.5)

a factor of two to five (at maximum loading). The difference between Cu concentration and activity in the forest floor solutions remained practically the same over the whole range of added copper. In the B_{sh} horizons this difference decreased with increasing Cu concentration in the equilibrium solutions.

Metal sorption was always accompanied by proton release to the solutions due to exchange reactions with H^+ ions from the soil matrix. However, in some mineral horizons pH did not change or even increased slightly, particularly at an initial pH of 3; these cannot be explained by proton release and may result from the net release of other cations such as Ca and Al. In addition, adsorption of the sulphate counterion to iron and aluminium oxides can contribute to proton consumption and pH increase; it is expected to be especially important at lower pH.

Correlations between DOC and pH were observed for all the soils; higher DOC was associated with lower pH, probably due to the solubilisation of organic matter sorbed to the soil via bridging cations (cf. de Wit *et al.*, 2001; this chapter). The most clearly defined correlations were revealed for the O horizons of K-200 (Fig. 5.8); also significant relationships between DOC and pH were found in some other soils (*r* from -0.59 to -0.77, α=0.001), though combining different soils and versions of the experiments led to appreciable

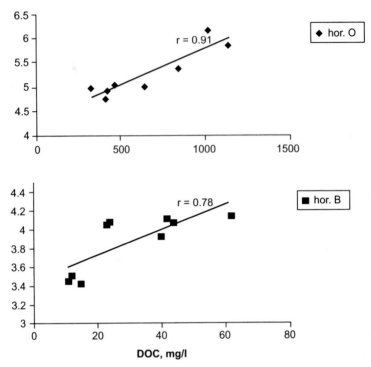

Fig. 5.7 Copper activity in the equilibrium solutions depending on DOC concentration (Cu_{in} = 1 mM).

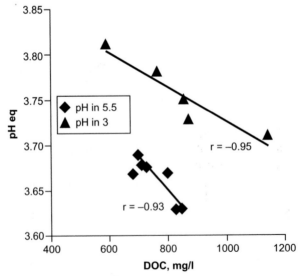

Fig. 5.8 Equilibrium DOC and pH in the soil solutions (K-200, hor. O, Cu+Ni)

increases in the variation of pH. Depending on the experimental conditions (soil, pH and metals concentrations in the input solutions) the decrease in pH could be very significant (up to two pH units). In turn, solution pH influenced metal sorption by proton competition.

Observations of equilibrium pH, total dissolved and free ionic metal may be explained by first considering a theoretical system where soil organic matter does not pass into solution (i.e. no DOC). The pH of the soil solution is determined by the acidity properties of the soil solids (predominantly organic matter), and the pools of chemically active ions (predominantly cations, e.g. Ca, Al). On addition of the metal salt, the system shifts to a new equilibrium state where the added metal is partially sorbed by the soil, with concomitant release of bound protons and other cations. The pH of the equilibrium solution is determined by the buffering capacity of the solid phase and the affinities of protons and other ions for the solids; as the added metal salts are Lewis acids a decrease in pH with increasing metal addition would be expected. The free metal ions, and the concentrations of dissolved inorganic complexes of the metal, are determined by the pH, the sorption capacity and the concentrations and relative affinities for the solid phase of competing ions. If we now consider the dissolution of soil organic matter (and associated bound metal) to form DOC, and assume to a first approximation that dissolution does not affect the chemistry of the organic matter, then the degree of dissolution and the DOC concentration do not affect the solution pH, since organic matter influences the chemistry of the system identically whether in the solid phase or dissolved. The actual DOC concentration is dependent upon many factors of which the most important is likely to be the total soil organic matter; therefore higher DOC concentrations (and consequently higher dissolved metals) are observed in more organic horizons. The higher DOC observed in the forest floor solutions (up to 100 times greater than in the illuvial horizons) promotes the dissolution of complexed metal resulting in the greater differences between total and free ion than in the illuvial horizons. Based on the higher binding affinity for Cu for organic matter, we would expect greater differences between dissolved and free Cu than between dissolved and free Ni.

The effect of pH on Cu and Ni sorption becomes most obvious at pH 3 (when competition from protons for binding sites is greatest). This occurs in the polluted soils (plots S-16 and S-41 are slightly polluted by Ni) and (or) when both Cu and Ni are present in the solutions.

The impact of pH on Ni and Cu sorption can be considered using soil from the K-200 plot as an example. This effect was investigated by considering the relationship between pH and (i) the parameters of the Langmuir and Henry equations; (ii) the amounts of sorbed Cu and Ni [Q_{Cu} and Q_{Ni} (Table 5.5)].

O-horizons. When $NiSO_4$ was added singly (Table 5.4, soil K-200), there was negligible effect of pH on $Q_{max, Ni}$ and $K_{H, Ni}$ (Figure 5.9). In the competition experiment, slight changes of $K_{H, Ni}$ were found: $K_{H, Ni} = 34\text{-}35$ at pH 4.0-5.5 and $K_{H, Ni} = 29$ at pH 3.0. The same trend was seen in the other soils (Fig. 5.9, Table 5.5). Ni adsorption by O-horizons could also be described by Freundlich and Langmuir equations (Table 5.4). Sorption parameters also decreased with decreasing pH: the maximum adsorption capacity $Q_{Ni, max}$ decreased between 20% and 30% with the pH decrease, in soil K-200. Values of Freundlich constant S_0 for Ni – at 11 (S-16)-26% (S-41). $Q_{max, Ni}$ in competition experiments decreased between 10% and 15% with the drop in pH.

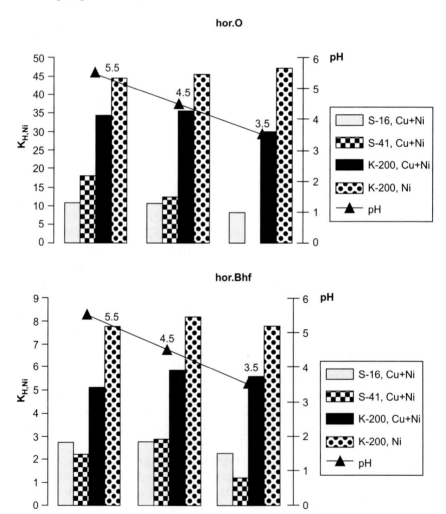

Fig. 5.9 Values of Henry constant, $K_{H,Ni}$ depending on the pH of the input solutions

The effect of pH on Cu sorption was considered for the competition experiments (Fig. 5.10). pH affected Cu sorption most noticeably in the humus horizons A (P-60) and peat horizons T (S-41) . When pH decreased from 5.5 to 3.0 $K_{H,Cu}$ decreased by a factor of between 100 and 400. $K_{H,Cu}$ in the O-horizon did not change by a factor of more than 50 (Table 5.5). This decrease is most prominent in the K-200 and S-41 soils. Soil S-16, closest to the smelter, showed a different response to changing pH. Soil S-16 is slightly polluted by Ni, and therefore the 25% decrease in Ni sorption with lower pH was more evident than the 10% decrease in Cu sorption.

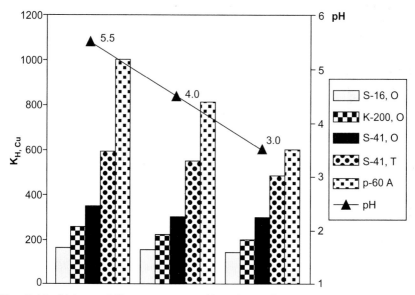

Fig. 5.10   Values of Henry constant, K_{Cu}, depending on the pH of the input solutions (Cu+Ni, hor. O)

B_{sh} horizon. The parameters of the Langmuir equations $Q_{max,Ni}$ correlate better with pH than do values of Henry constant $K_{H,Ni}$. Values of $K_{H,Ni}$ change irregularly both in the experiments with Ni only and with Cu and Ni together. The maximum adsorption capacity $Q_{max,Ni}$ decreases significantly with decreasing pH in most cases (by a factor of between 3 and 15 as pH dropped from 5.5 to 3.0, depending on the specific soil).

Adsorption of Cu by B_{sh} horizons is described only by Langmuir and Freundlich equations. In contrast to Ni, decreasing pH from 5.5 to 4.5 had no significant effect on $Q_{max,Cu}$ values, but values of k_{Cu} (the "affinity" constant) and $Q_{max,Cu}$ (at the maximum loadings) decreased significantly when the pH was decreased from 4.5 to 3.5. Freundlich parameters demonstrated irregular trends with changing pH.

Multiple regression analysis allows estimating the simultaneous effect of DOC and pH on the copper activity (a_{Cu2+}) in soil solutions. The following equations were generated for copper:

$$pa_{Cu\,eq} = 1.22\,pCu_{eq} + 1.0\,pH_{eq} - 1.6\,pDOC - 1.4 \text{ (for organic horizons)}$$
$$pa_{Cu\,eq} = 1.20\,pCu_{eq} - 0.39\,pDOC + 0.5 \text{ (for mineral horizons)}$$

The coefficient values show that the influence of pH and DOC on copper activity is greater in the organic horizons. Copper activity decreases with increasing DOC concentration and increases at reduced pH. The possibilities of the application of calculated equations in prediction copper activity in soils solutions was estimated by comparing two data sets - calculated and measured Cu activity values (Fig. 5.11). It was found that relative error of $pa_{Cu\,eq}$ calculated from the regression equations for O and B$_{sh}$ horizons did not exceed 12%, while the A and Ò soil horizons showed distinct deviations from the derived relationship.

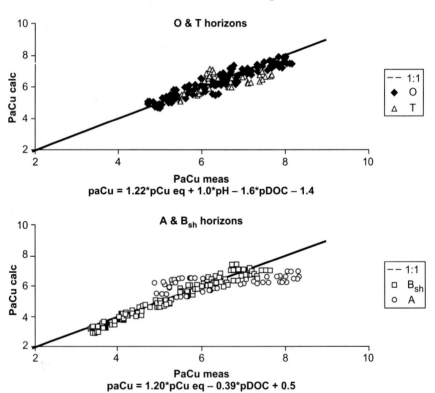

Fig. 5.11 Calculated versus measured paCu in organic (top) and mineral (bottom) horizons

Response of Soil Bacterial Communities to Heavy Metals Pollution

Bacterial counts

The numbers and biomass of soil bacteria decreased significantly along the gradient of HM pollution, with statistically significant changes close to the source of pollution (7-20 km from the smelter) (Table 5.6). The number of CFU per gram dry soil also decreased with decreasing distance from the smelter, as well as the ratio of CFU to total viable bacterial cell counts. Bacterial cell shape did not differ significantly along the distance from the smelter, whereas the average cell volume decreased with increasing distance (Table 5.6).

Discriminant analysis of the microscopic and agar plate counts distinguished 3 groups in the plane of the first 2 canonical variables: 7 km, 20-28 km, and 100-200 km (Fig. 5.12).

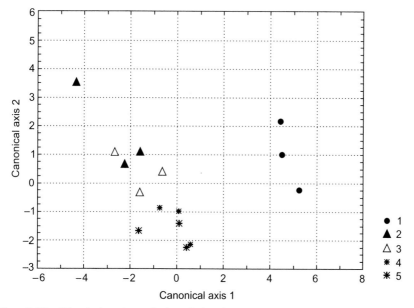

Fig. 5.12 Discriminant analysis of the bacterial community characteristics obtained microscopically: location of study sites in the plane of the first two canonical axes. Distance from the smelter: 1 – 7 km, 2 – 20 km, 3 – 28 km, 4 – 100 km and 5 – 200 km from the smelter.

Analysis of variance components showed that distance was major variance component in such bacterial community characteristics as the number of CFU (95%, $P<0.001$), the ratio of CFU to the total viable cell

Table 5.6 Some characteristics of soil bacterial communities in ecosystems along the gradient of heavy metals pollution

	Distance from the smelter and vegetation type					
	2 km, industrial barren	7 km, sparse forest with barren	20 km, sparse forest	28 km, defoliating forest	100 km, initial stage of defoliation	200 km, background forest
The number of cells in g o.d. soil, log	8.09 ± 0.19 a	8.80 ± 0.19 ab	9.29 ± 0.07 b	9.20 ± 0.16 b	9.33 b	9.34 ± 0.11 b
Bacterial biomass, $\mu g\ C \cdot g^{-1}$ o.d. soil	26 ± 19 a	42 ± 14 a	218 ± 35 b	176 ± 53 b	190 ± 66 b	201 ± 54 b
Cell volume, μm^3	0.390 ± 0.15 c	0.174 ± 0.007 b	0.170 ± 0.008 b	0.155 ± 0.03 b	0.141 ± 0.002 a	0.135 ± 0.02 a
Cell length-to-width ratio	1.45 ± 0.20 a	1.42 ± 0.05 a	1.42 ± 0.04 a	1.44 ± 0.04 a	1.49 ± 0.04 a	1.53 ± 0.02 a
The frequency of dividing cells	1.03 ± 0.54 a	3.91 ± 0.23 b	3.28 ± 0.41 b	2.89 ± 0.59 b	3.71 ± 0.95 b	3.32 ± 0.10 b
The number of CFU** in 1 g o.d. soil, log	3.44 ± 0.04 a	4.09 ± 0.45 a	5.50 ± 0.02 b	8.05 ± 0.08 c	7.89 ± 0.17 c	8.29 ± 0.18 c
The ratio of CFU to cell numbers, log	-4.65 ± 0.18 a	-4.71 ± 0.56 a	-3.79 ± 0.08 a	-1.15 ± 0.16 b	-1.38 ± 0.32 b	-1.05 ± 0.08 b

* The numbers in lines followed by different letters, differ significantly at P<0.05 level.

** Colony-forming units

number (90%, P=0.001) , average bacterial cell volume (80%, P=0.001) and bacterial biomass C (50%, P=0.05), while contributing much less the total number of viable cells (20%, P=0.21) and cell length-to-width ratio (16%, P=0.01) and having no influence on the frequency of dividing cells (P=0.75).

Community level physiological profiling

Figure 5.13 shows the relation between average well colour development and bacterial cell concentration in a Biolog Ecoplate well. The curves for the 100-200 km sites lie mostly above the other curves, with the curve for 7 km lying below all other curves and the curves for 20-28 km sites lying mainly in between (with occasional points lying even above the points for 100-200 km on the graph). The curve for the 2 km site lies above the 7-km curve, close to the 20-28 km curves.

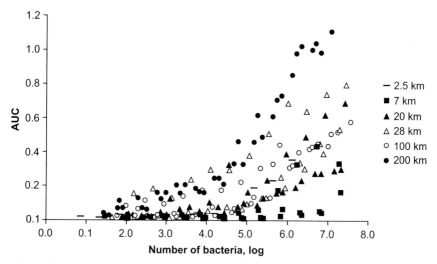

Fig. 5.13 Relationship between average well colour development (expressed as area under curve, AUC) and bacterial cell density in an Ecolog microplate well. Distance from the smelter: 1 – 7 km, 2 – 20 km, 3 – 28 km, 4 – 100 km and 5 – 200 km from the smelter.

Estimated EC_{50} values for the average well colour development showed a clear gradient, i.e. EC_{50} gradually decreased with distance from 7 km to 200 km (Table 5.7). However, the variation of the data also decreased, indicating correlation between the EC_{50} estimate and its variance. Interestingly, the industrial barren site fell out of the gradient, its EC_{50} value being the same as the EC_{50} value for the 100 km site.

Extraction of the principal components from the matrix with $LogEC_{50}$ values for different substrates as columns and the study sites as cases

Table 5.7 Estimation of the logarithm of effective cell concentration in a Biolog Ecoplate as averaged over all substrates

	2 km, industrial barren	7 km, sparse forest with barren	20 km, sparse forest	28 km, defoliating forest	100 km, initial stage of defoliation	200 km, background forest
			Distance from the smelter and ecosystem type			
LOG EC$_{50}$ Estimate	7.2	8.8	7.7	7.5	7.2	7.0
SEM	0.2	0.7	0.3	0.1	0.1	0.0
95% confidence intervals	6.77 ÷ 7.58	7.40 ÷ 10.20	7.22 ÷ 8.26	7.20 ÷ 7.74	6.97 ÷ 7.36	6.91 ÷ 7.11
EC$_{50}$ (cell well^{-1})	1.48E+07	6.30E+08	5.46E+07	2.95E+07	1.46E+07	1.02E+07

yielded 8 factors, accounting together for 92% of the matrix data variance, with the first 2 factors accounting for 37% and 15%, respectively. Denoting extracted factors as CLPP1, CLPP2... CLPP8, according to their contribution into the variance of the original data, we used these new variables for further statistical analyses. Here we will not interpret the variables, just stress that CLPP1 represents the decrease in community metabolic activity and diversity. CLPP1 correlated positively with average cell volume and negatively with CFU number and its ratio to direct counts (Table 5.8).

Table 5.8   Correlation coefficients between metabolic and other characteristics of soil bacterial communities

	Metabolic variables			
	CLPP1	CLPP4	CLPP6	CLPP8
Number of bacterial cells	−0.15	0.01	0.07	0.16
Bacterial biomass	−0.19	−0.07	0.07	0.33
Cell volume	0.68*	−0.21	0.08	0.02
Cell length-to-width ratio	−0.21	0.26	−0.10	−0.08
Frequency of dividing cells	−0.09	−0.03	0.03	−0.63*
Number of CFU**	−0.62*	−0.14	−0.05	0.30
Ratio of CFU to cell number	−0.64*	−0.16	−0.06	0.29

* – Marked correlations are significant at P<0.05 level

Discriminant analyses of the CLPP data clearly distinguished the 7 km site from all the others in the plane of the first two canonical axes (Fig. 5.14).

ANOVA analysis showed the distance from the smelter contributed from 15 to 20% to the total variance of CLPP1, CLPP4, CLPP6 and CLPP8 variables, i.e. the same variables that correlated with canonical roots in the discriminate analysis (results not shown). However, the distance effect was not statistically significant, the probability level ranging from 0.25 to 0.95. The main variance component was the interaction between distance and replication of soil samples, taken as an error term to test the significance of the distance and replication.

Variance components analysis of the $LogEC_{50}$ values of the individual substrates in the microplate showed that distance was contributing to the variance of half of the substrates of the microplate (Table 5.9), although the effect was not statistically significant. The substrates that displayed the maximal contribution of the distance factor into their variance at close to statistically significant levels (P<0.1) were phenylalanine, galacturonic acid and cyclodextrine.

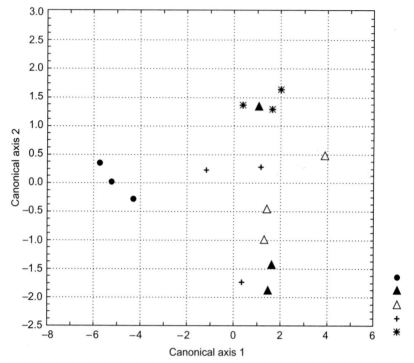

Fig. 5.14 Discriminant analysis of the bacterial community characteristics, obtained by factor analysis of the $LogEC_{50}$ data from the CLPP analysis: location of study sites in the plane of the first two canonical axes. Distance from the smelter: 1 – 7 km, 2 – 20 km, 3 – 28 km, 4 – 100 km and 5 – 200 km from the smelter.

Correlation between the bacterial and soil chemical variables

Most of the quantitative characteristics of the bacterial communities were strongly correlated with pH, the content and activity of Ni and Cu ions and field soil moisture (Table 5.10). Among the microscopic characteristics, cell volume was positively correlated with heavy metal content in soil and activity in soil solution, and negatively correlated with pH and water content. The length-to-width ratio of bacterial cells was positively correlated with pH only, whereas the frequency of dividing cells showed no correlation with soil chemical characteristics (Table 5.10). The metabolic variable CLPP1, which accounted for 36% of the total variance of the Log EC_{50} variance, was positively correlated with pH and heavy metals activity in soil solution, but not heavy metals content. Most of the other CLPP variables showed no statistically significant correlation (Table 5.10). However, the CLPP8, accounting for just 4% of the total $logEC_{50}$ variance, negatively correlated with both heavy metal ions activity in soil solution.

Table 5.9 ANOVA results for the log EC_{50} values of different substrates utilization in Biolog Ecoplates by bacterial communities from soils along the gradient of heavy metals pollution

Well	Substrate	Estimated relative variance (%)		
		Distance	Replication	Probability level for distance
C4	L-phenylalanine	43.4	4.4	0.06
B3	D-galacturonic acid	37.4	12.1	0.07
E1	α-cyclodextrine	36.4	4.2	0.09
F1	glycogen	22.0	13.3	0.18
A3	D-galactonic acid γ-lactone	20.8	0	0.22
B1	pyruvic acid methyl ester	19.0	0	0.24
E3	γ-hydroxybutiric acid	18.2	17	0.21
B2	D-xylose	16.6	0	0.26
E4	L-threonine	14.8	25.5	0.23
B4	L-asparagine	14.4	4.5	0.28
D1	Tween 80	13.4	5.2	0.29
A2	β-methyl-D-glucoside	10.6	0	0.33
D4	L-serine	9.9	2.7	0.33
H1	α-D-lactose	9.8	0	0.34
F3	itaconic acid	7.2	0	0.37
G4	phenylethylamine	1.9	0	0.44

Table 5.10 Correlation coefficients between soil chemical and bacterial characteristics

Bacterial characteristics	Soil chracteristics					
	pH_{CaCl_2}	Log Ni	Log aNi	Log Cu	Log aCu	Water content
Number of bacterial cells	0.60*	−0.49	−0.66**	−0.56*	−0.66**	0.76**
Bacterial biomass	0.60*	−0.56*	−0.75**	−0.67**	−0.78**	0.80***
Cell volume	−0.60*	0.67**	0.63*	0.60*	0.58 *	−0.58*
Cell length-to-width ratio	0.56*	−0.37	−0.46	−0.28	−0.38	0.44
Frequency of dividing cells	0.13	0.12	0.10	0.23	0.19	0.04
Number of CFU**	0.52*	−0.71**	−0.71**	−0.68**	−0.70**	0.58*
Ratio of CFU to cell number	0.46	−0.67**	−0.64*	−0.63*	−0.63*	0.49
CLPP1	−0.56*	0.50	0.58*	0.41	0.51*	−0.36
CLPP4	0.36	0.13	0.00	0.13	0.02	−0.14
CLPP6	−0.02	0.26	0.16	0.22	0.11	−0.22
CLPP8	0.14	−0.41	−0.52*	−0.49	−0.54*	0.37

*, ** and *** – Marked correlations are significant at P<0.05, P<0.01 and P<0.001 levels, respectively

The influence on zonal ecosystems of the long-term aerial deposition of heavy metals with the Severonikel smelter wastes decreases with increasing distance from the smelter. Along the distance range of 2-200 km from the smelter, ecosystems vary from an industrial barren to a practically undisturbed site, forming a sharp gradient of plant communities and soil characteristics. Thus the distance from the smelter represents a certain rate of aerial deposition of heavy metals, which, from the bacterial community point of view, is a complex factor, affecting all ecosystems components and exerting its influence on the soil microbial communities both directly, via the presence of metal ions in the soil solution and matrix, and indirectly, via other ecosystem components and soil physicochemical properties. Whatever the directly influencing factor, soil bacterial communities, being in close relation with soil solid, liquid and gaseous phases and other biological components, over more than 30 years of aerial contamination should develop distinct quantitative, morphological, physiological and metabolic characteristics at the sites located at different distances from the smelter along the aerial transportation of heavy metals.

Different quantitative characteristics of bacterial communities displayed different patterns of significant shifts between the soil ecosystems along the gradient of heavy metals deposition. Bacterial numbers and biomass both distinguished between 2-7 and 20-200 km ranges, while the number of the colony-forming units distinguished between 2-7, 20 and 28-200 km range (Table 5.6). Besides, the relative change in microscopic counts of bacteria along the gradient of heavy metals deposition was smaller compared to the change in the proportion of the community that formed colonies on the laboratory media (Table 5.6). Distance accounted for most of the variation in the CFU numbers, whereas it contributed much less to bacterial cell counts and biomass estimates. Thus plate counts seem to be a more appropriate method for determining the effect of heavy metals on soil bacterial communities than microscopic fluorescence counts, since the total numbers and biomass are highly sensitive to the quantity and quality of SOM and OM input with plant material. The proportion of bacteria from the soil samples that were culturable on standard plate-counting media varied between 0.01 and 9.3% and distinguished between 2-20 and 28-200 km ranges of distance from the smelter (Table 5.6), being negatively correlated with the content and activity of Ni and Cu in the soils (Table 5.10). This effect has been observed before (Roane and Pepper, 2000; Ellis *et al.*, 2003). So culturability needs to be taken into account while studying the effect of heavy metals on bacterial communities (Johnsen *et al.*, 2003), although the mechanisms of the effects are not that apparent since colonies on agar plates grow from one viable cell or spore, a group of cells or spores, and their combinations.

The average cell shape was not dependent on the rate of aerial pollution, thus not discriminating between the sites along the gradient, whereas the frequency of dividing cells dropped drastically in the industrial barren soil as compared to the rest of the studied distance range. However, the estimate of bacterial cell volume seemed to be a better indicator of the effect of heavy metals on soil bacteria, showing significant decrease from 2 km, to 7-28 km and then 100-200 km ranges (Table 5.6). It should be noted, though, that the variation of the average bacterial cell volume estimate for the industrial barren site was much higher than the same variation in other sites, indicating increased variation between soil replicates due to methodical difficulties of counting just a few bacterial cells per slide in this case. Since average cell volume is taken into account while estimating bacterial biomass from direct cell counts, bacterial biomass estimate is more indicative of the long-term effects of heavy metals than cell counts.

The concept of the effective concentration of cells implies that the bigger this parameter, the less active is a bacterial community in utilizing a certain substrate. The increase in the $logEC_{50}$ value, averaged for the 31 various substrates of an Ecoplate, with decreasing distance (Table 5.7) and hence the increasing rate of aerial heavy metals contamination, indicated decreasing metabolic activity of soil bacteria, and principally agrees with results from other studies (Knight et al., 1997; Bååtth et al., 1998; Shi et al., 2002; Jun Dai et al., 2004). Thus smaller bacterial cells in the almost uncontaminated ecosystems (Table 5.6) were more active metabolically. Higher metabolic activity of smaller cells may be a general phenomenon in soil ecosystems (De Fede and Sexstone, 2001).

Despite the fact that heavy metal free ion activity integrates soil chemical variability and that strong correlations of bacterial characteristics and heavy metals content and activity in soil was observed in our study (Table 5.10), agreeing with the data of other researchers (Dumestre et al., 1999), the nature of the effect of heavy metals on bacterial communities remains unclear, i.e. it is not known whether the effect is mostly direct or indirect. However, since microscopic and plating characteristics of bacteria showed much stronger influences of distance compared to metabolic characteristics and changed mostly within the distance range having sparse and defoliating forests (7-28 km), we believe that the effect is mostly indirect, primarily via quantitative and qualitative changes in plant material input. The importance of environmental factors, such as differences in C input, for microbial functioning in soil under long-term influence of heavy metals contamination, has been shown elsewhere (Chander et al., 2001) and explains why fresh additions of heavy metals do not model long-term effects, at least on microbial biomass and activity (Renella et al., 2002).

The correlation between metabolic bacterial variable and soil pH confirms the inhibiting effect of soil pH on microbial functioning (Knight et al., 1997; Blagodatskaya and Anderson, 1999).

In general our results agree well with results of researchers who employed similar analyses, both bacterial and statistical (Knight *et al.*, 1997).

However, the direct effect of increased concentrations of heavy metals in soils on bacterial functioning is indirectly supported by much stronger correlation between bacterial characteristics and metal ions activity in soil solution, rather then the total content (Table 5.10), and the absence of a statistically significant correlation between the bacterial metabolic variable and the total content of heavy metals, whereas the respective correlation with metal ion activity was highly significant. Other researchers (Evdokimova and Mozgova, 2003) showed that restoration of the number and activity of soil microorganisms at this site was associated with a reduction in soil metal toxicity.

The fact that the bacterial community of the industrial barren displayed $logEC_{50}$ values much lower than the respective values for the 7-28 km sites and hence higher metabolic activity is very interesting. Agar plating of bacterial suspensions from this soil showed active growth of actinomycetes (data not shown) on the plates, which is quite natural since the soil at this site is enriched with dead wood debris, with practically no addition of fresh plant material. We believe that metabolically the community has adapted to the very specific soil environment of the industrial barren, where heavy metal ion activity in soil solution is not mitigated by freshly added low-molecular weight organic substrates with rhizodeposition (Nigam *et al.*, 2001) and alike. So its metabolic activity is stabilized at a somewhat higher level, compared to the 7 km site. Revegetation of this site with azonal grasses for over 5 years resulted in a significant increase in the community-level metabolic profiles (Naumova *et al.*, 2004), once more emphasizing the importance of fresh plant C input for microbial metabolic activity and diversity.

Little research is devoted to sampling strategies to assess the effects of environmental and anthropogenic factors on microbial functional diversity in soils (Degens and Vojvodic-Vukovic, 1999; Balser, 2002), and most of the replication, if any, is quasi-replication at the level of a microplate or a subsample. We used individual soil replication, but due to obvious economic reasons adhered to the minimum number, i.e. we sampled three individual soil samples at each study site, subjecting them to further analyses. We used this statistical replication as a random factor for ANOVA. Its contribution to the total variance was practically always negligible; however, the influence of its interaction with distance factor was a major variance component for all the studied characteristics of bacterial communities, except for the CFU number and its ratio to viable cell counts. This interaction factor accounted for most of the variance of the CLPP variables and individual substrate utilization (Table 5.8). We consider the lack of the statistical significance in the effect of the distance factor on the metabolic characteristics of bacterial communities in our data to be due to the minimal field replication; in other words, three individual soil samples are clearly not enough to statistically demonstrate the effects of the long-term heavy metal deposition on forest soil ecosystems. A similar problem of high variation of Biolog data between soil replicates in studies of long-term heavy metal contamination of soils has been encountered before (Bååth *et al.*, 1998; Speir *et al.*, 2003), also resulting in no significant differences between contaminated and uncontaminated soils.

With this in mind, we believe that in reality there is a clear effect of heavy metal deposition on the utilization of at least such substrates as phenylalanine, galacturonic acid and cyclodextrine. Phenyl derivatives are known to be among the main products of plant material decomposition in podzols (Rumpel *et al.*, 2002), galacturonic acids are main components of plant cell walls pectins, and cyclodextrines are glucose oligomers. All these support our conclusion that the metabolic diversity of the bacterial community in the studied ecosystems is mainly governed by the changes in plant community, namely its production and the input of readily available C substrates.

Furthermore, we compared metabolic profiles of bacterial communities on the basis of equivalent cell number in a microplate well, thus actually addressing the properties of bacterial communities, rather than soil properties. It may also decrease the effect of heavy metal deposition since bacteria are known for their metabolic plasticity.

A significant effect of root exudates on the development of bacterial populations, especially culturable ones, in soil contaminated with heavy metals, has been shown before (Kozdroj and van Elsa, 2000; Ellis *et al.*, 2003). Within the distance range of 7-20 km, where vegetation is patchy and unstable, and this effect may be partly responsible for higher variation in bacterial characteristics, especially metabolic ones.

Some authors argue that, for practical monitoring purposes, microbial metabolic diversity may provide additional sensitivity in detecting effects of land management practices on the properties of soil microbial communities (Degens *et al.*, 2001). Fully agreeing with it, we should like to emphasize the importance of a sampling strategy and a methodical approach, and that special consideration should be given to the nature of the factor, for which microbes display this additional sensitivity, if any.

CONCLUSIONS

Thin podzols developed on unsorted and sorted till under scarce pine and spruce forests in the Kola Peninsula, Russia, are characterised by coarse texture, low content of organic carbon, low pH and low effective cation exchange capacity. Both natural (parent material, soil texture and mineralogy, organic matter) and man-induced (airborne pollution, vegetation disturbance) factors influence the current chemistry of podzols. Geological features of the territory and alkaline dust deposition prevent strong soil acidification near the pollution sources.

Atmospheric deposition of heavy metals from the nickel processing industry has led to severe contamination of surrounding soils within at least 100 km from the pollution sources. Concentrations of "total", nitric acid- and ammonium nitrate-extractable nickel and copper in organic horizons near the smelters were at least two orders of magnitude higher than the background levels in the region. The organic horizon of forest soils is an important accumulator of heavy metals and serves as a barrier against transport of pollution to underlying mineral horizons. Distribution of nickel and copper values in the B_{sh}- and C-horizons reflects mainly geological influences. However, the percentage of "total" metal content associated with mobile (acid-soluble and ammonium nitrate extractable) fractions in the mineral horizons increased near the smelter, probably indicating downward transport of deposited metal from the organic horizon.

In batch adsorption experiments in the pH range 3 to 5, soil horizons sorbed increasing amounts of Cu and Ni with increasing pH. O-horizons exhibited greater affinity for the metals than did B_{sh} horizons, due to their much higher SOM. Affinity for Cu was consistently greater than that for Ni. Free copper activity in the O-horizons was up to two orders of magnitude less than in the B_{sh} horizons, at similar total copper concentrations. However, under field conditions the copper activity in the B_{sh} horizons is expected to be lower than in the O horizons due to the much lower total copper concentrations; thus the environmental risk to soil communities is expected to be greatest in the organic horizons. Similar patterns in free Ni ion are expected although data were not available.

Taking to the account pH and DOC concentration in the equilibrium solutions and using multiple regression equations, values of $pa_{Cu\,eq}$ could be predicted. The relative error of the predictions was less than 12% for $pa_{Cu\,eq}$.

Under long-term contamination of forest ecosystems by aerial deposition of heavy metals from the smelter it is difficult to attribute conclusively the quantitative and qualitative changes in soil bacterial communities to direct effects of heavy metals. It is most likely that such effects, which can be observed within 20-30 km from the source of contamination, are mostly caused by drastic changes in plant components along the gradient, which occur within the same range, and strongly influence the availability of organic carbon or resources to heterotrophic bacteria. Direct effects of heavy metals on the bacterial community function are indicated by its significant correlation with metal free ion activity. Marked variations of the metabolic characteristics of bacterial communities among field replicates of soil samples may impede drawing conclusions about the effects.

ACKNOWLEDGEMENTS

We would like to express our appreciation to Dr. N. Lukina for her support and to the staff of the Analytical Laboratory of INEP KSC RAS for the AAS measurements. Special thanks are due to Zh. Vologdina for her help in the execution of the batch adsorption experiments and to S. Livantsova and I. Smirnova for chemical analyses. The study was supported by INTAS (01–2213) and the Russian Foundation of Basic Research (02-04-49047 and 02-04-48396à).

REFERENCES

Appelo, C.A.J., Postma, D. (1993). Geochemistry, groundwater and pollution. Rotterdam: A.A. Balkema.

Ashmore, M., Colgan, A., Farago, M., Fowler, D., Hall, J., Hill, M., Jordan, C., Lawlor, A., Lofts, S., Nemitz, E., Pan, G., Paton, G., Rieuwerts, J., Thornton, I., Tipping, E. (2001). Development of a critical load methodology for toxic metals in soils and surface waters: Stage II. Report to UK Department of the Environment, Transport and the Regions.

Bååth, E., Diaz-Ravina, M., Frostegard, A., Campbell, C.D. (1998). Effect of metal-rich sludge amendments on the soil microbial community. Applied and Environmental Microbiology 64: 238–245.

Balser, T.C., Kirchner, J.W., Firestone, M.K. (2002). Methodological variability in microbial community level physiological profiles. Soil Science Society of America Journal 66: 519-523.

Beining, B.A., Otte, M.L. (1996). Retention of metals originating from an abandoned lead-zinc mine by a wetland at Glendalough. Proceedings of the Royal Irish Academy, Vol 96B, No. 2, p. 117-126.

Benedetti, M.F., van Riemsdijk, W.H., Koopal, L.K. (1996). Humic substances considered as a heterogeneous donnan gel phase. Environmental Science and Technology: 30:1805-1813.

Blagodatskaya, E.V., Anderson, T.H. (1999). Adaptive responses of soil microbial communities under experimental acid stress in controlled laboratory studies. Applied Soil Ecology 11: 207-216.

Bloem, J. (1995). Fluorescent staining of microbes for total direct counts. In Molecular Microbial Ecology Manual, pp. 1-12. Eds A.D.L. Akkermans, J.D. van Elsas and F.J. de Brujin. Kluwer Publishers, Amsterdam.

Bloem, J., Veninga, M., Shepherd, J. (1997). Fully automated measurement of soil bacteria by confocal laser scanning microscopy and image analysis. Scientific and Technical Information XI: 143-148.

Bloem, P.R. (1981). Metal-organic matter interactions in soil. In: Chemistry in the soil environment, pp. 129-150. Eds. R.H. Dowdy et al. ASA Spec. Publ. 40. ASA and SSSA, Madison, WI.

Bloomfield, C. (1953). A study of podzolization. Part 1. Journal of Soil Science 4: 5-16.

Bloomfield, C.(1954). A study of podzolization. Parts 3-5. Journal of Soil Science 5: 39-56.

Chairidchai, P., Ritchie, G.S.P. (1990). Zn adsorption by lateric soil in the presence of organic ligands. Soil Science Society of America Journal 54: 1242-1248.

Chander, K., Dyckmans, J., Joergensen, R.G., Meyer, B., Raubuch, M. (2001). Different sources of heavy metals and their long-term effects on soil microbial properties. Biology and Fertility of Soils 34: 241–247.

De Fede, K.L., Sexstone, A.J. (2001). Differential response of size-fractionated soil bacteria in Biolog™ microtitre plates. Soil Biology and Biochemistry 33: 1547-1554.

De Vries, W., Bakker, D.J. (1998). Manual for calculating critical loads of heavy metals for terrestrial ecosystems. Guidelines for critical limits, calculation methods and input data. Wageningen, The Netherlands: DLO Winand Staring Centre Report 166.

de Wit, H.A., Groseth, T., Mulder, J. (2001). Predicting aluminum and soil organic matter solubility using the mechanistic equilibrium model WHAM. Soil Science Society of America Journal 65: 1089-1100.

Degens, B.P., Schipper, L.A., Sparling, G.P., Duncan, L.C. (2001). Is the microbial community in a soil with reduced catabolic diversity less resistant to stress or disturbance? Soil Biology and Biochemistry 22: 1143-1153.

Degens, B.P., Vojvodic-Vukovic, M. (1999). A sampling strategy to assess the effects of land use on microbial functional diversity in soils. Australian Journal of Soil Research 37: 593-601.

Dijkstra, F.A., Geibe, C., Holmstrom, S., Van Breeme, N. (2001). The effect of organic acids on base cation leaching from the forest floor under six north American tree species. European Journal of Soil Science 52: 205-214.

Dumestre, A., Sauvé, S., McBride, M., Baveye, P., Berthelin, J. (1999). Copper speciation and microbial activity in long-term contaminated soils. Archives of Environmental Contamination and Toxicology 36: 124–131.

Edwards, M., Benjamin, M.M., Ryan, J.N. (1996). Role of organic acidity in sorption of natural organic matter (NOM) to oxide surfaces. Coll. Surf. A – Physicochem. Eng. Asp. 107: 297-307.

Elliot, H.A., Denneny, C.M. (1982). Soil adsorption of Cd from solutions containing organic ligands. Journal of Environmental Quality 11: 658-663.

Ellis, R.J., Morgan, P., Weightman, A.J., Fry, J.C. (2003). Cultivation-dependent and -independent approaches for determining bacterial diversity in heavy-metal-contaminated soil. Applied and Environmental Microbiology 69: 3223-3230.

Evdokimova, G.A., Mozgova, N.P. (2003). Restoration of properties of cultivated soils polluted by copper and nickel. Journal of Environmental Monitoring 5: 667-670.

Fotovat, A., Naidu, R. (1998). Changes in composition of soil aqueous phase influence chemistry of indigenous heavy metals in alkaline sodic and acidic soils. Geoderma 84: 213-234.

Gallet, C., Keller, C. (1999). Phenolic composition of soil solutions: comparative study of lysimeter and centrifuge waters. Soil Biology and biochemistry 31: 1151-1160.

Garland, J.L. (1997). Analysis and interpretation of community-level physiological profiles in microbial ecology. FEMS Microbiology and Ecology 24: 289–300.

Garland, J.L., Mills, A.L., Young, J.S. (2001). Relative effectiveness of kinetic analysis vs single point readings for classifying environmental samples based on community-level physiological profiles (CLPP). Soil Biology and Biochemistry 33: 1059-1066.

Haas, C.I., Horowitz, N.D. (1986). Adsorption of Cd to kaolinite in the presence of organic materials. Water, Air and Soil Pollution 27: 131-140.

Houba, V.J.G., van der Lee, J.J., Walinga, I., Novozamsky, I. (1985). Soil analysis, Part 2: Procedures. Department of Soil Science and Plant Nutrition, Wageningen Agricultural University, The Netherlands.

Inskeep, W.P., Baham, J. (1983). Competitive complexation of Cd(11) and Cu(11) by water-soluble organic ligands and Na-montmorillonite. Soil Science Society of America Journal 47: 1109-1115.

Jardine, P.M., Weber, N.L., McCarthy, J.F. (1989). Mechanisms of dissolved organic-carbon adsorption on soil. Soil Science Society of America Journal 53(5): 1378-1385.

Johnsen. K., Ekelund, F., Binnerup, S.J., Rasmussen, L.D. (2003). Mercury decreases culturability of *Pseudomonas frederiksbergensis* JAJ 28 in soil microcosms. Current Microbiology 47: 125-128.

Johnson, C.E., Romanowicz, R.B., Siccama, T.G. (1997). Conservation of exchangeable cations after clear-cutting of northern hardwood forest. Canadian Journal of Forest Research 27: 859-868.

Johnson, C.E., Siccama, T.G., Driscoll, C.T., Likens, G.E., Moeller, R.E. (1995). Changes in lead biogeochemistry in response to decreasing atmospheric inputs. Ecological Applications 5: 813-822.

Jun, Dai., Thierry Becquer, T., Rouiller, J.H., Reversat, G., Bernhard-Reversat, F., Lavelle, P. (2004). Influence of heavy metals on C and N mineralisation and microbial biomass in Zn-, Pb-, Cu-, and Cd-contaminated soils Applied Soil Ecology 25: 99–109.

Kaiser, K., Guggenberger, G., Haumaier, L., Zech, W. (2001). Seasonal variations in the chemical composition of dissolved organic matter in organic forest floor layer leachates of old-growth Scots pine (Pinus sylvestris L.) and European beech (Fagus sylvatica L.) stands in northeastern Bavaria, Germany. Biogeochemistry 55: 103–143.

Kaiser, K., Guggenberger, G., Zech, W. (1996). Sorption of DOM and DOM fractions to forest soils. Geoderma 74(3-4): 281-303.

Kaiser, K., Zech, W. (1997). Competitive sorption of dissolved organic matter to soils and related mineral phases. Soil Science Society of America Journal 61(1): 64-69.

Kaiser, K., Zech, W. (1999). Release of natural organic matter sorbed to oxides and subsoil. Soil Science Society of America Journal 63: 1157-1166.

Karczewska, A. (1999). Mobilization of HM from polluted soils as affected by pH and other factors. Proceedings of the 5th International Conference on the Biogeochemistry of trace elements. V. 6. Vienna, pp. 624-625.

Kashulina, G., Reimann, C., Finne, T.E., Halleraker, J.H., Äyräs, M., Chekushin, V.A. (1997). The state of the ecosystems in the central Barents Region: scale, factors and mechanism of disturbance. The Science of the Total Environment 206: 203-225.

Kinniburgh, D.G., Milne, C.J., Benedetti, M.F., Pinheiro, J.P., Filius, J., Koopal, L.K., VanRiemsdijk, W. H. (1996). Metal ion binding by humic acid: Application of the NICA-Donnan model. Environmental Science and Technology 30(5): 1687-1698.

Knight, B.P., McGrath, S.P., Chaudhari, A.M. (1997). Biomass carbon measurements and substrate utilization patterns of microbial populations from soils amended with cadmium, copper, or zinc. Applied and Environmental Microbiology 63: 39–43.

Konopka, A., Oliver, L., Turco, R.F. Jr. (1998). The use of carbon substrate utilization patterns in environmental and ecological microbiology. Microbial Ecology 35: 103–115.

Koopal, L.K., Van Riemsdijk, W.H., de Wit, J.C.M., Benedetti, M.F. (1994). Analytical isotherm equations for multicomponent adsorption to heterogeneous surfaces. Journal of Colloid and Interface Science 166: 51–160.

Koptsik, G.N., Niedbaiev, N.P., Koptsik, S.V., Pavlyuk, I.N. (1998). Heavy metal pollution of forest soils by atmospheric emissions of Pechenganikel smelter. Eurasian Soil Science 31(8): 896-903.

Koptsik, G., Koptsik, S., Venn, K., Aamlid, D., Strand, L., Zhuravleva, M. (1999). Changes in acidity and cation exchangeable properties of forest soils under atmospheric acid deposition. Eurasian Soil Science 32 (7): 787-798.

Koptsik, S., Koptsik, G., Livantsova, S., Eruslankina, L., Zhmelkova, T., Vologdina, Zh. (2003). Heavy metals in soils near the nickel smelter: chemistry, spatial variation, and impacts on plant diversity. Journal of Environmental Monitoring 5: 441–450.

Kozdroj, J., van Elsas, J.D. (2000). Response of the bacterial community to root exudates in soil polluted with heavy metals assessed by molecular and cultural approaches. Soil Biology and Biochemistry 32: 1405-1417.

Leenheer, J.A. (1981). Comprehensive approach to preparative isolation and fractionation of dissolved organic carbon from natural waters and waste waters. Environmental Science and Technology 15: 578-587.

Lock, K., Janssen, C.R. (2001). Modeling zinc toxicity for terrestrial invertebrates. Environmental Toxicology and Chemistry 20(9): 1901–1908.

Lofts, S., Spurgeon, D.J., Vendsen, C., Tipping, E. (2004). Deriving soil critical limits for Cu, Zn, Cd, and Pb: A method based on free ion concentrations. Environmental Science and Technology 38: 3623-3631.

Lukina, N.V., Nikonov, V.V. (1996). Acidity and chemical composition of soil water in Al-Fe-humus podzolic soils under pine forests. Eurasian Soil Science 29: 196–205.

Lukina, N., Nikonov, V. (2001). Assessment of environmental impact zones in the Kola Peninsula forest ecosystems. Chemosphere 42: 19-34.

Lundström, U.S. (1993). The role of organic acids in the soil solution chemistry of a podzolized soil. Journal of Soil Science 44: 121-133.

Malcolm, R.L., McCracken, R.J. (1968). Canopy drip: a source of mobile soil organic matter for mobilization of Fe and Al. Soil Science Society of America Proceedings 32: 834-838.

Mazet, M., Angbo, L., Serpaud, B. (1990). Adsorption of humic acids onto performed aluminium hydroxide flocs. Water Resources 24: 1509-1518.

Michalzik, B., Matzner, E. (1999). Dynamics of dissolved organic nitrogen and carbon in a Central European Norway spruce ecosystem. European Journal of Soil Science 50: 579-590.

Naidu, R., Harter, R.D. (1998). Effect of different organic ligands on cadmium sorption by and extractability from soils. Soil Science Society of America Proceedings 62: 644-651.

Naumova, N.B., Rutgers, M., Lukina, N. (2004). Metabolic diversity of soil bacterial community under revegetation of an industrial barren, resulting from the long-term heavy metals pollution. Sibirsky Ecologichesky Journal 3: 455-464 (in Russian).

Neal, R.H., Sposito, G. (1986). Effect of soluble organic matter and sewage sludge amendments on Cd sorption by soils at low Cd concentrations. Soil Science 142: 164-172.

Nierop, K.G.J., Jansen, B., Vrugt, J.A., Verstraten, J.M. (2002). Copper complexation by dissolved organic matter and uncertainty assessment of their stability constants. Chemospere 49: 1191-1200.

Nigam, R., Srivastava, S., Prakash, S., Srivastava, M.M. (2001). Cadmium mobilisation and plant availability – the impact of organic acids commonly exuded from roots. Plant and Soil 230: 107–113.

Nikonov, V., Goryainova, V., Lukina, N. (2001a). Ni and Cu migration and accumulation in forest ecosystems on the Kola Peninsula. Chemosphere 42: 93-100.

Nikonov, V.V., Lukina, N.V., Polyanskaya, L.M., Panikova, A.N. (2001b). Distribution of microorganisms in the Al-Fe-Humus Podzols of natural and anthropogenically impacted boreal spruce forests. Mikrobiologiya 70(3): 374-383.

Ong, H.L., Bisque, R.E. (1968). Coagulation of humic colloids by metal ions. Soil Science 106: 220-224.

Oste, L.A., Temminghoff, E.J.M., van Riemsdijk, W.H. (2002). Solid–solution partitioning of organic matter in soils as influenced by an increase in pH or Ca concentration. Environmental Science and Technology 36: 208–214.

Pennanen, T., Frostegard, A., Fritze, H., Bååth, E. (1996). Phospholipid fatty acid composition and heavy metal tolerance of soil microbial communities along two heavy metal-polluted gradients in coniferous forests. Applied and Environmental Microbiology 62: 420–428.

Pohlman, A.A., McColl, J.G. (1988). Soluble organics from forest litter and their role in metal dissolution. Soil Science Society of America Journal 52: 265-271.

Preston-Mafham, J., Boddy, L., Randerson, P.F. (2002). Analysis of microbial community functional diversity using sole-carbon-source utilisation profiles - a critique. FEMS Microbiology and Ecology 42: 1-14.

Reimann, Ñ., Äyräs, M., Chekushin, V., Bogatyrev, I., Boyd, R., Caritat, P de, Dutter, R., Finne, T.E., Halleraker, J.H., Jaeger, Ø., Kashulina, G., Niskavaara, H., Lehto, O., Pavlov, V., Räisänen, M.L., Srand, T., Volden, T. (1998). Environmental Geochemical Atlas of the Central Barents Region. NGU-GTK-CKE special publication: Geological Survey of Norway: Trondheim, Norway. 745 pp.

Renella, G., Chaudri, A.M., Brookes, P.C. (2002). Fresh addition of heavy metals do not model long-term effects on microbial biomass and activity. Soil Biology and Biochemistry 34: 121-124.

Roane, T.M., Pepper, I.L. (2000). Microbial responses to environmentally toxic cadmium. Microbial Ecology 38: 358-364.

Rumpel, C., Kögel-Knabner, L., Bruhn, F. (2002). Vertical distribution, age and chemical composition of organic carbon in two forest soils of different pedogenesis. Organic Geochemistry 33: 1131-1142.

Sauvé, S., McBride, M., Hendershot, W. (1998a). Soil solution speciation of lead: effects of organic matter and pH. Soil Science Society of America Journal 62: 618-621.

Sauvé, S., Dumestre, A., McBride, M., Hendershot, W. (1998b). Derivation of soil quality criteria using predicted chemical speciation of Pb^{2+} and Cu^{2+}. Environmental Toxicology and Chemistry 17: 1481-1489.

Schnitzer, M., Scinner, S.I.M. (1967). Organo-metallic interactions in soils: 7. Stability constants of Pb^{2+}, Ni^{2+}, Mn^{2+}, Co^{2+}, Ca^{2+} and Mg^{2+} fulvic acid complexes. Soil Science 103: 247-252.

Shi, W., Bischoff, M., Turco, R., Konopka, A. (2002). Long-term effects of chromium and lead upon the activity of soil microbial communities. Applied Soil Ecology 21: 169–177.

Speir, T.W., van Schaik, A.P., Lloyd-Jones, A.R., Kettles, H.A. (2003). Temporal response of soil biochemical properties in a pastoral soil after cultivation following high application rates of undigested sewage sludge. Biology and Fertility of Soils 38: 377–385.

Spurgeon, D.J., Hopkin, S.P. (1996). Effects of variations of the organic matter content and pH of soils on the availability and toxicity of zinc to the earthworm Eisenia fetida. Pedobiologia 40: 80–96.

Stevenson, F.J. (1994). Humus chemistry: Genesis, composition, reactions. 2nd edition. John Wiley & Sons. New York.

Stevenson, F.J., Welch, L.F. (1979). Migration of applied lead in a field soil. Environmental Science and Technology 13: 1255-1259.

Strobel, B.W. (2001). Influence of vegetation on low molecular weight carboxylic acids in soil solution – a review. Geoderma 99: 169-198.

Temminghoff, E.J.M., Van der Zee, S.E.A.T.M., De Haan, F.A.M. (1998). Effect of dissolved organic matter on the mobility of Cu in a contaminated soil. European Journal of Soil Science 49: 617-628.

Tipping, E., Woof, C. (1990). Humic substances in acid organic soils: modelling their release to the soil solution in terms of humic charge. Journal of Soil Science 41: 573–586.

Tipping, E. (1994). WHAM – a chemical speciation program and computer code for waters, sediments, and soils incorporating a discrete site/electrostatic model of ion-binding by humic substances. Computational Geosciences 20: 973–1023.

van Hees, P.A.W., Lundstrom, U.S., Giesler, R. (2000). Low molecular weight acids and their Al complexes in soil solution- composition, distribution, seasonal variation in three podzolized soils. Geoderma 94: 173-200.

Vologdina Zh, V., Koptsik, G.N., Karavanova, E.I. (2004). Main features of copper sorption by podzols of Kola Peninsula. Moscow University Soil Science Bulletin (accepted).

Vorob'eva, L.A. (1998). Chemical Analysis of Soils. Textbook. Moscow: Moscow University Press. 272 pp.

Zysset, M., Berggren, D. (2001). Retention and release of dissolved organic matter in Podzol B horizons. European Journal of Soil Science 52(3): 409-421.

CHAPTER 6

Bioremediation

Andrew S. Ball[*]

Department of Biological Sciences, University of Essex, Wivenhoe Park,
Colchester CO4 3SQ, UK
[*]Present Address: School of Biological Sciences, Flinders University,
GPO Box 2100, Adelaide SA 5001, Australia.

Bioremediation can broadly be described as the use of living organisms to degrade environmental pollution or to prevent pollution through waste treatment. Bioremediation must compete with existing methods in terms of efficiency and economy. However, the biotechnological solution to the problem requires only moderate capital investment, a low energy input, are environmentally safe, do not generate waste, and are self-sustaining. It is expected that future biotechnological methods of toxic waste treatment will play a key role both as a displacement for existing disposal methods and for the detoxification of novel xenobiotic (man-made) compounds. This chapter examines the techniques developed for bioremediatoion, including monitoring. The advantaged and limitations of bioremediation are also discussed. The final section looks at specific bioremediation examples and discusses the future of this rapidly growing environmental biotechnology.

INTRODUCTION

A general definition of the term bioremediation is "the use of microorganisms to return an object or area to a condition which is not harmful to plant or animal life". A useful example of bioremediation is the treatment of wastewater such as domestic sewage to reduce biochemical oxygen demand. This is a long-standing and well-understood example of using living organisms to return a material to its original state (Alexander, 1994). However, over the last two decades there has been a more specific use of the term bioremediation. This is reflected in the two more commonly used and more specific definitions:

- The use of living organisms to degrade environmental pollution or to prevent pollution through waste treatment (Atlas, 1995).
- The application of biological treatment to the cleanup of hazardous chemicals (Cookson, 1995).

ADVANTAGES OF BIOREMEDIATION

The bioprocesses for treating toxic effluents must compete with existing methods in terms of efficiency, economy, and safety. However, the biotechnological solution to the problem requires only moderate capital investment, a low energy input, is generally environmentally safe, it does not generate significant waste, and it is a self-sustaining process. It is expected that future biotechnological methods of toxic waste treatment will play a key role both as a displacement for existing disposal methods and for the detoxification of novel xenobiotic (man-made) compounds. Table 6.1 outlines the advantages that bioremediation has over other technologies (Cookson, 1995).

Table 6.1   Advantages of bioremediation

Some Other Advantages of Bioremediation
• Can be conducted on site
• Permanent elimination of waste (limiting liability)
• Positive public acceptance
• Minimum site disruption
• Eliminates transportation cost and liability
• Can be coupled with other treatment techniques
(Adapted from Cookson, 1995)

In many cases bioremediation not only offers a permanent solution to contamination problems but it is also cost effective. Table 6.2 compares the cost of various treatments commonly used for the removal of pollutants.

Table 6.2   A comparison of the costs of bioremediation with the costs of traditional methodologies

Method	Cost Effectiveness of Bioremediation ($)		
	Year 1	Year 2	Year 3
Incineration	530[1]	None	None
Solidification	115	None	None
Landfill	670	None	None
Thermal Desorption	200	None	None
Bioremediation	175	27	20

1 – costs are per cubic yard
(Adapted from Cookson, 1995)

Cleaning up existing environmental contamination in the United States alone is estimated to cost as much as $1 *trillion* dollars (Cookson, 1995).

Bioremediation can help reduce the costs of treatment as follows (Kerr, 1993):

Treating Contamination in Place

Most of the cost associated with traditional cleanup technologies is associated with physically removing and disposing contaminated soils. Because engineered bioremediation can be carried out in place by delivering nutrients to affected soils, it does not incur removal-disposal costs (Kerr, 1991).

Harnessing Natural Processes

At some sites, natural microbial processes can remove or contain contaminants without human intervention. In these cases where intrinsic bioremediation (natural attenuation) is appropriate, substantial cost savings can be realized (Kerr, 1993).

Reducing Environmental Stress

Because bioremediation methods minimize site disturbance compared with conventional cleanup technologies, post-cleanup costs can be substantially reduced (Kerr, 1993).

From the above discussion it can be concluded that bioremediation is generally more cost effective than either landfilling or incineration, the two other widely used methods for site cleanup (Cookson, 1995).

TECHNOLOGIES INVOLVED IN BIOREMEDIATION

Specific terms are used to describe the activity of the microorganisms and the way these organisms are used in site cleanup. It is important to firstly define these terms (Kerr, 1993; Bedient et al., 1994).

Biodegradation

The breaking down of a compound or substance is achieved with living organisms such as bacteria or fungi. These could be indigenous to the area, or could be introduced.

Biostimulation

The natural or introduced populations of microbes in an area are enhanced through addition of nutrients, engineering, or other manipulation of an affected site. This speeds the natural remediation process.

Bioaugmentation

Specific organisms are applied to a site or material to achieve a desired bioremediation effect.

Biorestoration

Restoration to the original or near-original state using microbes.

The technology used to treat a polluted site is dependent on the site and the pollutant. Therefore each bioremediation treatment is site-specific. However, the different treatments rely on a number of basic technologies which are described below :

In Situ Bioremediation (ISB)

ISB is the use of microorganisms to degrade contaminants in place with the goal of generating harmless chemicals as end-products. Most often, *in situ* bioremediation is applied to the degradation of contaminants in saturated soils and groundwater, although bioremediation in the unsaturated zone can also occur. The technology was developed as a less costly, more effective alternative to the standard pump-and-treat methods used to clean up aquifers and soils contaminated with chlorinated solvents, fuel hydrocarbons, explosives, nitrates, and toxic metals. ISB has the potential to provide advantages such as complete destruction of the contaminant(s), lower risk to site workers, and lower equipment and operating costs.

ISB can be categorized by mode of microbial metabolism or by the degree of human intervention. Microbial metabolism will be either aerobic or anaerobic. The target metabolism for an ISB system will depend on the contaminants of concern. Some contaminants (e.g., fuel hydrocarbons) are degraded via an aerobic pathway, some anaerobically (e.g., carbon tetrachloride), and some contaminants can be biodegraded under either aerobic or anaerobic conditions (e.g., trichloroethene) (Cantor *et al*. 1987; Kerr, 1993).

Accelerated In Situ Bioremediation

Where substrate or nutrients are added to an aquifer to stimulate the growth of a target consortium of bacteria. Usually the target bacteria are indigenous; however, enriched cultures of bacteria from other sites that are highly efficient at degrading a particular contaminant can be introduced into the aquifer (i.e., 'bioaugmentation'). Accelerated ISB is used where it is desired to increase the rate of contaminant biotransformation, which may be limited by lack of required nutrients, electron donor, or electron acceptor. The type of amendment required depends on the target metabolism for the contaminant of interest. Aerobic ISB may only require the addition of oxygen, while anaerobic ISB often requires the addition of both an electron donor (e.g., lactate, benzoate) as well as an electron acceptor (e.g., nitrate, sulfate). Chlorinated solvents, in particular, often require the addition of a carbon substrate to stimulate reductive dechlorination. The goal of accelerated ISB is to increase microbial biomass and activity throughout the contaminated volume of aquifer, thereby achieving effective biodegradation of dissolved and sorbed contaminants (Cantor *et al.*, 1987; Kerr, 1993).

Monitored Natural Attenuation (Intrinsic Bioremediation)

It is the second approach to applying *in situ* bioremediation. One component of natural attenuation is the use of indigenous microorganisms to degrade the contaminants of concern without human intervention (such as supplementing the available nutrients). Site characterization and long term monitoring comprise the activities required to implement natural attenuation. The site characterization determines the extent of contamination and the properties of the aquifer. This characterization information can then be used in a reactive transport model to predict the fate of contaminants and whether the contaminants will affect the receptors of concern. Long-term monitoring is used to assess the fate and transport of the contaminants compared against predictions. The reactive transport model can then be refined to obtain better predictions (Cantor *et al.*, 1987; Kerr, 1993).

Natural attenuation processes typically occur at all sites, but to varying degrees of effectiveness depending on the types and concentrations of contaminants present and the physical, chemical, and biological characteristics of the soil and groundwater. Natural attenuation processes may reduce the potential risk posed by site contaminants in three ways.

1. The contaminant may be converted to a less toxic form through destructive processes such as biodegradation or abiotic transformations.
2. Potential exposure levels may be reduced by lowering of concentration levels (through destructive processes, or by dilution or dispersion).
3. Contaminant mobility and bioavailability may be reduced by sorption to the soil matrix.

Whether accelerated ISB or natural attenuation is used at a particular site will depend upon the aquifer properties, chemical concentrations, goals of the remediation project, and the economics of each option. The rate of contaminant degradation is typically slower in a natural attenuation scenario than for active bioremediation because the concentration of soil microorganisms is much greater during accelerated bioremediation and the biodegradation rate is proportional to the amount of biomass. Thus, natural attenuation typically takes longer to complete. Accelerated ISB usually results in a more rapid solution, but requires a much greater investment in materials, equipment, and labour (Cantor *et al.*, 1987; Kerr, 1993).

Advantages of *In Situ* Bioremediation

- Contaminants may be completely transformed to innocuous substances (e.g., carbon dioxide, water, ethane).
- Accelerated ISB can provide volumetric treatment, thereby treating both dissolved and sorbed contaminant.
- The time required to treat subsurface pollution using *in situ* bioremediation can often be faster than pump-and-treat processes.
- *In situ* bioremediation often costs less than other remedial options.
- The zone of treatment using bioremediation can be larger than with other remedial technologies because the treatment moves with the plume and can reach areas that would otherwise be inaccessible.

Limitations of *In Situ* Bioremediation

- Depending on the particular site, some contaminants may not be completely transformed to innocuous products.
- If biotransformation reactions stall with the production of an intermediate compound, the intermediate may be more toxic and/or mobile than the parent compound.
- Some contaminants cannot be biodegraded (i.e., they are recalcitrant).
- When inappropriately applied, injection wells may become clogged from profuse microbial growth resulting from the addition of nutrients, electron donor, and/or electron acceptor.
- Accelerated *in situ* bioremediation is difficult to implement in low-permeability aquifers because transport of nutrients is limited.
- Heavy metals and toxic concentrations of organic compounds may inhibit activity of indigenous microorganisms.
- *In situ* bioremediation usually requires an acclimated population of microorganisms which may not develop for recent spills or for recalcitrant compounds.

Advantages of Monitored Natural Attenuation

- As with any *in situ* process, generation of lesser volume of remediation wastes, reduced potential for cross-media transfer of contaminants commonly associated with *ex situ* treatment, and reduced risk of human exposure to contaminated media.
- Less intrusion because few surface structures are required.

- Potential for application to all or part of a given site, depending on site conditions and cleanup objectives.
- Can be used in conjunction with, or as a follow-up to, other (active) remedial measures.
- Lower overall remediation costs than those associated with active remediation.

Disadvantages of Monitored Natural Attenuation

- Longer time frames may be required to achieve remediation objectives, compared to active remediation.
- Site characterization may be more complex and costly.
- Toxicity of transformation products may exceed that of the parent compound.
- Long term monitoring will generally be necessary.
- Institutional controls may be necessary to ensure long-term protection of the site and limit liability.
- Potential exists for continued contaminant migration, and/or cross-media transfer of contaminants.
- Hydrologic and geochemical conditions amenable to natural attenuation are likely to change over time and could result in renewed mobility of previously stabilized contaminants, adversely impacting remedial effectiveness.
- More extensive education and outreach efforts may be required in order to gain public acceptance of monitored natural attenuation.

APPLICATION OF BIOREMEDIATION

The rapid expansion and increasing sophistication of the chemical industries in the past century and particularly over the last thirty years has resulted in increasing volumes and complexity of toxic waste effluents. At the same time, fortunately, regulatory authorities have been paying more attention to environmental contamination. Industries are therefore becoming increasingly aware of the political, social, environmental and regulatory pressures to prevent escape of effluents into the environment. The occurrence of major incidents (such as the Exxon Valdez oil spill, the Union-Carbide (Dow) Bhopal disaster, large-scale contamination of the Rhine River, the progressive deterioration of the aquatic habitats and conifer forests in the Northeastern US, Canada, and parts of Europe, or the release of radioactive material in the Chernobyl accident) and the subsequent massive publicity due to the resulting environmental problems has highlighted the potential for imminent and long-term disasters in the public's conscience (Barcelona *et al.* 1990).

Even though policies and environmental efforts should continue to be directed towards applying pressure to industry to reduce toxic waste production, bioremediation presents opportunities to detoxify a whole range of industrial effluents (Figure 6.1).

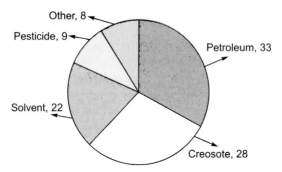

Fig. 6.1 The major types of waste chemicals that are applicable to bioremediation

Bacteria can be altered to produce certain enzymes that metabolize industrial waste components that are toxic to other life, and also new pathways can be designed for the biodegradation of various wastes. Since waste management itself is a well-established industry, genetics and enzymology can be simply "bolted-on" to existing engineering expertise. It can therefore be concluded that bioremediation is capable of treating a wide range of waste chemicals. As chlorinated hydrocarbons are one of the most common groups of compounds that require bioremediation technology, further examination of their degradation is outlined below (Hickey and Smith, 1996).

BIOREMEDIATION OF CHLORINATED HYDROCARBONS

Chlorinated hydrocarbons can undergo biotransformation via three different mechanisms: use of the chlorinated compound as an electron acceptor, use of the chlorinated compound as an electron donor, or by cometabolism (fortuitous reaction providing no benefit to the microorganisms). More than one of these mechanisms may be active at a given site (Hinchee *et al.*, 1994).

Electron Acceptor Reactions

Use of chlorinated compounds as electron acceptors has been demonstrated under nitrate- and iron-reducing conditions, but the most rapid biodegradation rates, affecting the widest range of chlorinated aliphatic hydrocarbons, occur under sulphate-reducing and methanogenic conditions. This mode of biotransformation requires an appropriate source of carbon (electron donor) for microbial growth and reductive dehalogenation to occur. The electron donor carbon source may include natural organic matter, anthropogenic sources (e.g., fuel hydrocarbon co-contamination), or intentional introduction of organic carbon into the aquifer (i.e., during accelerated *in situ* bioremediation) (Hinchee *et al.*, 1994).

Electron Donor Reactions

In this situation, the carbon source is used as the primary substrate (electron donor) and the microorganism obtains energy and organic carbon from the CS. This may occur under aerobic and some anaerobic conditions. Lesser oxidized chlorinated compounds (e.g., vinyl chloride, DCE, or 1,2-dichloroethane) are more likely to be amenable to this mode of biotransformation. Note that fuel hydrocarbons are biodegraded under this mode of metabolism because they can be used as an organic carbon source (Hinchee *et al.*, 1994).

Cometabolism

When a chlorinated aliphatic hydrocarbon is biodegraded via cometabolism, the degradation is catalyzed by an enzyme or cofactor that is fortuitously produced by the organisms for other purposes. The microbe receives no known benefit from the degradation of the chlorinated compound. The biotransformation of the CS may actually be harmful or inhibitory to the microorganism. Cometabolism is best documented in aerobic environments, although it potentially could occur under anaerobic conditions (Hinchee *et al.*, 1994).

The terms "reductive dehalogenation," "direct dechlorination," and "primary substrate" have been used elsewhere in an inconsistent manner, leading to potential confusion. The electron donor/acceptor terminology provides an explicit description. In general, reductive dechlorination is the process whereby a chlorine atom is removed from a compound and is replaced with a hydrogen atom. Direct dechlorination is usually associated with the chlorinated hydrocarbon acting as the electron donor. The primary substrate usually also refers to the electron donor.

MARKET PROSPECTS FOR BIOREMEDIATION

Over the last few years, a number of companies have been established to develop and commercialize biodegradation technologies. Existence of such companies now has become economically justifiable, because of burgeoning costs of traditional treatment technologies, increasing public resistance to such traditional technologies (ranging from Love Canal to the ENSCO incinerator plant in Mobile, Arizona some years ago), accompanied by increasingly stringent regulatory requirements. The interest of commercial businesses in utilizing microorganisms to detoxify effluents, soils, etc. is reflected in "bioremediation" having become a common buzzword in waste management. Companies specializing in bioremediation will need to develop a viable integration of microbiology and systems engineering. As an example of a bioremediation company, Envirogen (NJ) has developed recombinant PCB (polychlorinated biphenyl)-degrading microorganisms with improved stability and survivability in mixed populations of soil organisms. The same company also has developed a naturally occurring bacterium that degrades trichloroethylene (TCE) in the presence of toluene, an organic solvent which is toxic to many microorganisms. A wide range of similar companies are now in existence. Microorganisms were also successfully applied during cleanup activities at the Exxon Valdez oil spill of 1989. A number of microorganisms can utilize oil as a source of food, and many of them produce potent surface-active compounds that emulsify oil in water and facilitate the removal of the oil. Unlike chemical surfactants, the microbial emulsifier is non-toxic and biodegradable. Also, simple inorganic fertilizers have been utilized to increase the growth rate of the indigenous population of bacteria that are able to degrade oil.

Use of microbes for bioremediation is not limited to detoxification of organic compounds. In many cases, selected microbes can also reduce the toxic cations of heavy metals and metalloids (such as selenium) to the much less toxic and less soluble elemental form. Thus, bioremediation of surface water with significant contamination by heavy metals can now be carried out. Figure 6.2 summarises the potential size of this industry by examining the number of polluted sites that have been identified in the US that require clean up (Ministry of Environment and Energy, 1992; US Department of Commerce, 1993).

EXAMPLE OF THE SUCCESSFUL APPLICATION OF BIOREMEDIATION

The problem in Hanahan, South Carolina, a quiet suburb of Charleston, was not particularly unusual. In 1975, a massive leak from a military fuel storage facility released about 80,000 gallons of kerosene-based jet fuel. Immediate and extensive recovery measures managed to contain the spill, but could not prevent some fuel from percolating into the permeable sandy soil and reaching the underlying water table. Soon, ground water was leaching toxic

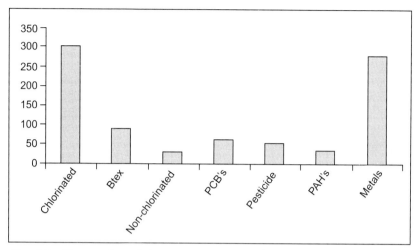

Fig. 6.2   The number of sites in the USA that require treatment for pollution

chemicals such as benzene from the fuel-saturated soils and carrying them toward a nearby residential area.

By 1985, contamination had reached the residential area and the facility was faced with a serious environmental and public health problem. Removing the contaminated soils was technically impractical, and removing contaminated ground water did not address the source of the contaminants. How could contaminated groundwater be kept from seeping toward the residential area?

Studies by the U.S. Geological Survey (USGS) had shown that microorganisms naturally present in the soils were actively consuming fuel-derived toxic compounds and transforming them into harmless carbon dioxide. Furthermore, these studies had shown that the rate of these biotransformations could be greatly increased by the addition of nutrients. By stimulating the natural microbial community through nutrient addition, it was theoretically possible to increase rates of biodegradation and thereby shield the residential area from further contamination.

In 1992, this theory was put into practice. Nutrients were delivered to contaminated soils and substrata through infiltration galleries, contaminated ground water was removed by a series of extraction wells, and the arduous task of monitoring contamination levels began. By the end of 1993, contamination in the residential area had been reduced by 75 per cent. Nearer to the infiltration galleries (the source of the nutrients), the results were even more promising. Ground water that once had contained more than 5,000 parts per billion toluene now contained no detectable contamination (EPA region 5, 1995a, b).

The success of the Hanahan Bioremediation Project was no accident. It was the result of many years of intensive effort by many scientists and engineers.

In the early 1980's, little was known about how toxic wastes interact with the hydrosphere. This lack of knowledge was crippling efforts to remediate environmental contamination under the new U.S. Superfund legislation—the Comprehensive Environmental Response, Compensation, and Liability Act. Faced with this problem, Congress directed the USGS to conduct a program to provide this critically needed information. By means of the Toxic Substances Hydrology Program, the most important categories of wastes were systematically investigated at sites throughout the United States. One of the principal findings of this program was that microorganisms in shallow aquifers affect the fate and transport of virtually all kinds of toxic substances. For example:

Crude Oil Spill, Bemidji, Minnesota

In 1979, a pipeline carrying crude oil burst and contaminated the underlying aquifer. USGS scientists studying the site found that toxic chemicals leaching from the oil were rapidly degraded by natural microbial populations. Significantly, it was shown that the plume of contaminated groundwater stopped spreading after a few years as rates of microbial degradation came into balance with rates of contaminant leaching. This was the first and best-documented example of intrinsic bioremediation in which naturally occurring microbial processes remediated contaminated ground water without human intervention.

Sewage Effluent, Cape Cod, Massachusetts

Disposal of sewage effluent in septic drain fields is a common practice in non-urban areas throughout the United States. Systematic studies of a sewage effluent plume at Massachusetts Military Reservation (formerly known as Otis Air Force Base) led to the first accurate field and laboratory measurements of how rapidly natural microbial populations degrade nitrate contamination (denitrification) in a shallow aquifer.

Chlorinated Solvents, New Jersey

Chlorinated solvents are a particularly common contaminant in the heavily industrialized Northeastern United States. Because their metabolic processes are so adaptable, microorganisms can use chlorinated compounds as oxidants when other oxidants are not available. Such transformations, which can naturally remediate solvent contamination of groundwater, has been extensively documented by USGS scientists at Picatinny Arsenal, New Jersey.

Pesticides, San Francisco Bay Estuary

Pesticide contamination of rivers and streams is a matter of concern throughout the United States. Field and laboratory studies in the Sacramento River and San Francisco Bay have shown the effects of biological and non-biological processes in degrading commonly used pesticides, such as molinate, thiobencarb, carbofuran and methyl parathion.

Agricultural Chemicals in the Mid-continent

Agricultural chemicals affect the chemical quality of groundwater in many Midwestern States. Studies in the midcontinent have traced the fate of nitrogen fertilizers and pesticides in ground and surface waters. These studies have shown that many common contaminants, such as the herbicide atrazine, are degraded by biological (microbial degradation) and non-biological (photolytic degradation) processes.

Gasoline Contamination, Galloway, New Jersey

Gasoline is probably the most common contaminant of groundwater in the United States. Studies at this site have demonstrated rapid microbial degradation of gasoline contaminants and have shown the importance of biological processes in the unsaturated zone (the zone above the water table) in degrading contaminants.

Creosote Contaminants, Pensacola, Florida

Creosote and chlorinated phenols have been used extensively as wood preservatives throughout the United States. Contaminants leaked to the underlying aquifer through several unlined ponds and were transported toward nearby Pensacola Bay. Studies at this site have demonstrated that microorganisms can adapt to extremely harsh chemical conditions and that microbial degradation was restricting migration of the contaminant plume.

Together, these studies laid the technical foundation that enabled bioremediation to be applied at Hanahan.

Further Reading

A web address with links to sites related to bioremediation: http://www.nal.usda.gov/bic/Biorem/biorem.htm. As is apparent from this site, the US government, in particular the US Department of Energy, has a keen interest in bioremediation (for example, see the NABIR (Natural and Accelerated Bioremediation Research) Program at http://www.lbl.gov/NABIR/index.html). Part of this interest stems from the commitment to clean up heavily polluted sites (such as the Hanford site in Washington state) that once served as nuclear weapons facilities. A US DOE website on Hanford cleanup, involving both traditional technologies and bioremediation, can be found at http://www.hanford.gov/rl/index.asp.

REFERENCES

Alexander, M. (1994). Biodegradation and bioremediation. Academic Press, San Diego.

Atlas, R.M. (1995). Bioremediation. Chemical and Engineering News 73 (14): 32-42.

Barcelona, M., Keely, J., Pettyjohn, W., Wehrmann, A. (1988). Groundwater protection. Hemisphere Publishing Corporation.

Barcelona, M., Wehrmann, A., Keely, J., Pettyjohn, W. (1990). Contamination of groundwater, Noyes Data Corp., New Jersey, pp. 35-47, 87-107.

Bedient, P., Rifari, H.C., Newell, C.P. (1994). Groundwater contamination, Prentice-Hall, Inc. New Jersey.

Cantor L., Knox, R., Fairchild, D. (1987). Groundwater quality protection, Lewis Publishing Inc., Michigan.

Cookson, J.T. (1995). Bioremediation engineering design and application. McGraw Hill, New York.

EPA Region 5. (1995a). Admiral petroleum in BFSS, Greenville, MI.

EPA Region 5. (1995b). Wesco Oil #37 in BFSS, Muskegon, MI.

Hickey, R., Smith, G. (1996). Biotechnology in industrial waste and treatment and bioremediation. Lewis Publishers, London.

Hinchee, R.E., Alleman, B.C., Hoeppel, R.E., Miller, R.N. (1994). Hydrocarbon bioremediation. Lewis Publishers, London.

Kerr, R.S. (1993). In-Situ bioremediation of ground water and geological material: A Review of Technologies. US-EPA, U.S.A.

Ministry of Environment and Energy. (1992). Remediation technologies for contaminated soils, Ontario, Canada.

Kerr, R.S. (1991). Characterization for subsurface remediation. US-EPA, U.S.A.

U.S. Department of Commerce. (1993) Water quality standards handbook 2nd Edition. USA.

CHAPTER 7

Effects of Fly Ash on Soil Characteristics, Plant Growth and Soil Microbial Populations

Zaki A. Siddiqui and Lamabam P. Singh

Department of Botany, Aligarh Muslim University, Aligarh 202 002 INDIA.

Utilization of fly ash is a practical and important waste management issue, and agronomic uses of fly ash are being explored as a means of disposal. Fly ash changes the physico-chemical properties of soil. Ash applications increase soil pH and electrical conductivity. Fly ash contains numerous plant nutrients and improves the nutritional status of soil. It also contains heavy metals and other compounds which are toxic to plants and microbes. Fly ash changes microbial populations and negatively effects cycling of nutrients. Over-application of fly ash can cause toxicity to animals grazing on recipient soils. Various plant pathogens show variable response to their host plants in fly ash- amended soil. Extensive field trials are required for making proper recommendations of doses of fly ash amendments to soil.

INTRODUCTION

The worldwide growth in thermal power generation capacity does not come without its share of problems. Coal combustion produces huge quantities of fly ash as a by-product, which often accumulates at the facility. Portions of this waste may be blown by winds, thus becoming an additional environmental hazard. Currently, 90 million tones of fly ash are generated annually in India. With a planned increase in coal-based power plants in the coming year this figure is expected to reach 200 million tones by 2012. Already 65,000 acres of land is used for ponds where fly ash is dumped (Srivastava, 2002). Moreover, only about 12 million tones or 13 per cent of the fly ash generated is being utilized in India (Srivastava, 2002), but in advanced countries, the utilization rate is about 65 per cent of generation (Sengupta, 2002).

Fly ash has been used in civil construction, building materials, land and mine fill, in agriculture and land management. It has been established beyond doubt that fly ash can be advantageously used in agriculture as an agent for acidic soils, and as a conditioner for improving some soil physico-chemical properties such as hydraulic conductivity, bulk density, porosity, water holding capacity, etc. Fly ash has also been used as a source of essential plant nutrients like calcium, magnesium, potassium, phosphorus, copper, zinc, manganese, iron, boron, and molybdenum, and ultimately for boosting crop growth and yield. Fly ash has been applied in different agro-climatic conditions and soil types in different parts of the world at rates ranging from 25-500 t ha^{-1} (Sengupta, 2002).

Field trials with fly ash amendments have shown appreciable increases in yield for different crops with high nutritional value, i.e., protein and oil, without any significant carryover of trace / heavy metals and radioactivity to soil, water bodies, or crop produce. Fly ash has the potential to alter soil texture by increasing the percentage of silt-sized particles, and thus permanently increasing soil moisture holding capacity (Jones and Amos, 1976). Bulk density is often increased, especially in mine soils. Some fly ash materials, particularly those from sub-bituminous coals, can have a cementing effect (pozzolonic activity), which is controlled primarily by the CaO content of the ash. Indian fly ash, being generally alkaline in nature with pozzolonic properties, inhibits the infiltration rate in sandy (loose textured) soils and enhances the infiltration rate of clay-rich soil due to the higher proportion of silt–sized particles in the ash. Fly ash has also been successfully used in bulk scale for reclamation of degraded lands and low-lying areas for growing various crops and other plant species. Reclamation of abandoned ash ponds through biological means via planta-tions has been established at many thermal power plants. Bulk use of fly ash in agriculture or forestry may well be successful if it can substitute for chemical fertilizers.

PHYSICAL PROPERTIES OF FLY ASH

Fly ash generally has a silt loam texture with 65-90% of particles measuring less than 0.010 mm in diameter (Chang et al., 1977; Roy et al., 1981). Ash is an amorphous glassy material composed of silica, alumina and iron oxides, with other metals present in smaller quantities. The constituents, apart from the glass, that are of most significance to the mechanical properties of fly ash are calcium oxide (lime) and sulfate. Ash from bituminous coal is usually finer than that produced by the burning of lignite (Tolle et al., 1982). In general, fly ash has low bulk density (1.01-1.43 g cm^{-3}), and hydraulic conductivity (Roy et al., 1981; Tolle et al., 1982; Mattigod et al., 1990).

CHEMICAL PROPERTIES OF FLY ASH

The chemical composition of fly ash depends on the composition of the coal, combustion conditions, the efficiency and type of emission control devices and the disposal methods used (van Hook, 1979; Adriano *et al.*, 1980). Major elemental components of fly ash are Al, Fe and Si with smaller concentrations of Ca, K, Na, Ti and S. Several ashes have high Ca concentrations because of surface CaO deposits. Ash from bituminous coals is generally higher in Fe, K and S, and lower in Mg and Ca than that from sub-bituminous and lignitic coals (Theis and Wirth, 1977). Fly ashes contain varying amounts of trace elements. Some of these are required at low levels for plant and animal nutrition, but they may become toxic when present in higher concentrations. Other trace elements present in ash (such as Cd) play no known role in nutrition, and may be toxic to plants, animals and microbes if they become bioavailable (Daniels *et al.*, 2002).

When water is added to fly ash, initial pH values are low as the sulfate deposited on the particle surfaces is brought into solution as sulphuric acid. This is a transient situation and after a short time, the pH rapidly rises as calcium is leached into solution. The pH is typically 8 to 11 for fly ash, although the pH for ash with higher free calcium oxide contents can rise to 12. Only a small quantity of free calcium is required to achieve the higher pH (Daniel *et al.*, 2002).

The liming potential of fly ash is derived primarily from CaO, MgO and other alkaline metal oxides that react with water to generate net alkalinity. Ashes high in Ca and Mg oxides have a high neutralizing capacity (Daniels *et al.*, 2002). The property of fly ash which appears to correlate best with potential to produce alkalinity is its water-soluble calcium content, while acid-producing potential is best measured by its amorphous (oxalate-extractable) iron content. A rule of thumb for estimating the acidity or alkalinity of a given ash is that if the ratio of amorphous iron to water-soluble calcium in an ash is under 3.0, the ash will be alkaline (Theis and Wirth, 1977).

Inorganic Compounds

The inorganic composition of fly ash is dependent upon the mineral matter present within the coal that is fired in the furnace. Only a small fraction of the constituents are present on the surface of fly ash and are leachable in water. Major water-soluble constituents are calcium and sulfur (usually present as sulfate). There are smaller amounts of sodium and potassium, and traces of chloride, magnesium, aluminium and silicon (Sear *et al.*, 2003).

More than 70% of the Ti, Na, K, Mg, Hf, Th and Fe in ash are associated with the alumino-silicate matrix and more than 70% of the Ca, Sc, Sr, La with the rare earth elements. Nickel is apparently associated with an acid-soluble phase, which has the same particle size distribution as the alumino-silicate phases. More than 70% of the As, Se, Mo, Zn, Cd, W, V, and Sb were associated with surface material on the particles. The elements Mn, Be, Cr, Cu, Co, Ga, Ba and Pb were intermediate in behavior, i.e., distributed about equally between matrix and non-matrix material (Hansen and Fisher, 1981).

Incomplete combustion of coal may result in higher leachable trace metals from ash (Wu *et al.*, 1982). Fly ash will contain higher heavy metal concentrations compared with the original coal. Except in the case of Pb, Cd, and Co, heavy metals were more concentrated in the <53 μm than in the >150 μm fraction, suggesting the probable contribution of these metals to atmospheric emissions and waste ash. It is therefore of value to estimate the enrichment factors from coal to fly ash (Fulekar *et al.*, 1983). Metals in coal fly ash collected from electrostatic precipitators at thermal power plants are found to have monthly variations in particle size, per cent silicates, haemolysis and contents of Ni, Pb, Mn, Zn, Hg, Cd, Mg, Na, Fe, Ca, K, Cu, Sr, As, Se, Co and Cr (Srivastava *et al.*, 1986). Fly ash contains relatively moderate concentrations of Cd and Cu and much higher concentrations of Al, Pb and Zn, which were readily leached under normal landfill conditions (Hermann and Wiswewski, 1988; Sawell *et al.*, 1989). Ash samples contained marked amounts of trace elements including Cd and Pb and other metals such Co, Ca, Fe, Mn, Mo, Ni and Zn (Wong and Wong, 1989). Major matrix elements of fly ash were Si, Al, K, Fe, and Hg. Some of the biologically toxic

Table 7.1   Physico-chemical characteristics of soil mixed with fly ash

Characteristics	Fly ash in field soil, percent						C.D. P=0.05
	0	20	40	60	80	100	
C.E.C. (meq(100g)$^{-1}$)	4.2f	4.7e	5.2d	6.1c	8.1b	9.2a	0.4
Conductivity (mmhos cm^{-1})	3.3e	3.6e	4.6d	5.7c	7.2b	7.6a	0.3
pH	6.6d	6.8d	7.5c	7.8c	8.3b	8.9a	0.3
Porosity (%)	43.4d	43.7d	53.6c	59.4b	71.7a	74.5a	4.2
Water holding capacity (%)	39.2e	46.3d	55.4c	67.0b	83.2a	85.0a	4.3
Bicarbonate (%)	0.44f	0.71e	1.02d	1.44c	2.15b	2.61a	0.015
Carbonate (%)	0.20f	0.35e	0.42d	0.68c	0.89b	0.15a	0.022
Chloride (%)	0.26f	0.40e	0.63d	0.95c	1.32b	1.86a	0.016
Sulphate (%)	3.20f	4.50e	5.70d	6.90c	9.10b	9.74a	0.40
Nitrogen (%)	0.15f	0.12e	0.09d	0.07c	0.02b	0.00a	0.008
Phosphorus (%)	0.02f	0.02e	0.04d	0.05c	0.07b	0.09a	0.005
Potassium (%)	0.19f	0.22e	0.41d	0.47c	0.65b	0.80a	0.016
Calcium (%)	0.16f	0.26e	0.46d	0.71c	0.90b	1.05a	0.008

*Different letters within one character denote value that are significantly different at P=0.05

elements Ni, Cu, Pb, B and Mo were also present in substantial quantities. Ash from electrostatic precipitators was finer in texture, lower in pH, and richer in nutrients than ash collected from a dump site (Sikka *et al.*, 1994). Heavy metals such as Ni, As, Cd, Cr, Pb, Si, Zn and Cu (Wadge and Hutton, 1987; Onliveros *et al.*, 1990) and complex compounds viz., dibenzofurans and dibenzo-p-dioxins (Helder *et al.*, 1982; Sawyer *et al.*, 1983) are of particular concern with regard to toxicity to plants and soils.

Organic Compounds

Organic compounds within ash are influenced by the effectiveness of combustion and cooling conditions. Fly ash contains very little organic material other than elemental carbon. Combustion in the power station furnace, which has peak temperatures in excess of 1250°C will remove and break down most organic compounds present in the original coal. No synthetic organic compounds such as pesticides, for example, will be detected. There is usually a small amount of elemental carbon in fly ash, but nothing that will undergo organic degradation. Consequently, fly ash deposits do not produce gases from biodegradation reactions. The elemental carbon is partially activated and capable of absorbing some deleterious compounds such as polycyclic aromatic hydrocarbons (PAHs).

Polycyclic aromatic hydrocarbons

The most potentially hazardous PAHs are benzofluoranthenes, benzo [a] pyrene, benz [a] anthracene, dibenzo [a, h] anthracene and indenol [1,2,3-cd] pyrene. The total PAH concentration in fly ash has generally been found to be less than 0.5 mg kg^{-1}, due to the binding of PAHs to ash particles, which is less than 1% of the lower threshold of 50 mg kg^{-1} (Sear *et al.*, 2003). The potential strong binding of PAHs to fly ash, which is reflected in problems with extraction, is consistent with its behavior in soils. This binding dictates much of the environmental behaviour of PAHs, including their low availability for plant uptake and minimal groundwater leaching. PAHs are only sparingly soluble in water, with solubility decreasing with increasing size of the molecule. PAHs undergo photodegradation and are therefore thought to have a limited life span in the atmosphere. Naphthalene, phenanthrene, acenapthene and other compounds, which occur in greater proportions in fly ash, are in general the least toxic PAHs and subject to rapid adsorption and reasonably rapid degradation. The most toxic PAHs and the more slowly degrading substances are either not detected or are present at very low concentrations compared to naphthalene and similar PAHs. All PAHs adsorb readily and strongly onto other organic materials such as soils, and they are degraded with half-lives from a few days to a few years. Degradation is particularly rapid (half-life in hours) if the compounds are exposed to UV radiation from sunlight.

Phenols

The fate of phenols in ash is similar to that for PAHs except that the half-life of most phenols is considerably less than that for PAHs. They are present in very low concentrations that will be retained in ash deposits and released slowly, if at all. There will be some slow "in deposit" degradation and the minute amount of phenols which could be released will be rapidly adsorbed and degraded. Since the concentrations in the ash are undetectable and rapid adsorption or degradation will reduce concentrations, it is therefore inconceivable that phenols can cause significant soil contamination (Sear *et al.*, 2003).

Dioxins

Dioxins are usually associated with the incomplete combustion of material containing chlorine and as such are commonly associated with the ash from municipal waste incineration. Dioxin levels in fly ash are consistently very low, due to the low chlorine content of coal combined with the high temperatures found in the furnaces of power stations. Dioxins are ubiquitous and present in a wide range of soils, although they rapidly decay when exposed to light. Fly ash samples were found to contain polychlorinated dibenzo dioxins, dibenzofurans, and bromochloro-polynuclear aromatic hydrocarbons in ppt to ppb levels (Sovocool *et al.*, 1988).

Leachable Compounds Within Fly Ash

Fly ash is widely reported as having very low solubility. Most of the solid elemental inventory is held within the largely insoluble "glassy" alumino-silicate matrix. Typically, ~2% of fly ash is considered soluble and the eluates are dominated by calcium sulfate (gypsum), a naturally occurring compound found in soils, with lesser contributions from sodium, potassium and chloride ions. Most of the metals and metalloids present in ash are either retained in glass beads or firmly adhere to them, resulting in a very low leaching potential. Fly ash is also alkaline with a pH of typically 9-12, which materially assists in retaining metals (Sear *et al.*, 2003).

Many other compounds are believed to be present as oxides, sulfates and chlorides in the enriched surface layer of fly ash particles. These may be released and precipitated as secondary solids or adsorbed onto the solid substrate depending on their solubility limits.

Only a few elements are fully soluble in the initial leachate because of their limited solubility at high pH (Sear *et al.*, 2003). Other than calcium, the most mobile element is boron, which exists primarily as borate. Boron exists as a vapour at the temperature of molten ash, hence its availability in fly ash. When the ash solidifies, the boron condenses on its surface rather than being incorporated within the glass matrix. Weathering may eventually mobilize up to 50% of any boron and molybdenum together with 25-30% of the arsenic and selenium in the ash. The actual behaviour of these and other elements appears to be specific to the ash source and the total amount present (Sear *et al.*, 2003).

Release of trace elements from fly ash varies with the sample and is related to leaching solution, but it is not simply a function of soil pH. In sulfuric acid, barium dissolved from fly ash would probably form insoluble barium sulfate. Arsenic is most soluble in a basic solution and is only sparingly soluble in all other solutions. The average release of beryllium, cadmium, cobalt, lead, antimony and selenium is 2 mg kg^{-1} or less in all leaching solutions. Chromium, copper and nickel are most soluble in acetic and sulfuric acids, but the total amount leached is still less than 10 mg kg^{-1}. Zinc is also soluble in acetic and sulfuric acids, as well as in a ferric chloride solution (Kim and Kazonich, 2003). Polychlorinated dibenzo-p-dioxins and dibenzofurans (PCDD/Fs) from a fly ash column were eluted with pure water or an aqueous solution of linear alkyl benzene sulfonate [LAS]. Concentrations of PCDD/Fs in the leachates as well as relative leaching increased with increasing degree of chlorination (decreasing water solubility). Concentrations of LAS enhanced PCDD/Fs solubility (Schramm *et al.*, 1995).

EFFECTS OF FLY ASH ON SOIL

Due to its high concentration of various elements, numerous studies have evaluated the usefulness of fly ash in nutrient-deficient soil. It can be used as a source of K (Martens *et al.*, 1970), Mo (Martens *et al.*, 1970; Cary *et al.*, 1983; Elseewi and Page, 1984), B (Plank and Martens, 1974; Wallace and Wallace, 1986), Zn (Schnappinger *et al.*, 1975; Wallace *et al.*, 1980), Ca (Martens and Beahm, 1976; Wallace *et al.*, 1980), S (Elseewi *et al.*, 1978; Hill and Lamp, 1980), Cu (Wallace *et al.*, 1980) and Mg (Hill and Lamp, 1980). A beneficial use of fly ash on land could be as an amendment to mitigate low soil pH (<5) (Summer *et al.*, 1991). Deleterious effects of soil acidity include lower solubility of P, Ca, Mg, Zn and Cu which are essential to plants (Mengel and Kirkby, 1982); and greater solubility of many trace elements (As, Cd, Cr, Pb, Ni) which may be phytotoxic and detrimental to animals and humans when sufficient quantities of plant materials are consumed (Kabata-Pendias and Adriano, 1995). The pH of acidic soils usually needs to be increased to alleviate many detrimental effects such soils induce on plants. Addition of alkaline amendments will increase the pH of acidic soils (Plank *et al.*, 1975; Elseewi *et al.*, 1980; Moliner and Street, 1982; Elseewi and Page, 1984; Petruzzelli *et al.*, 1987; Khan *et al.*, 1996). The increase in pH depends primarily on soil buffering capacity; soils that are poorly buffered are expected to show the greatest increase in pH after ash treatment.

Soil is also amended with ash in order to correct micronutrient deficiencies. Acidic-to-neutral fly ash has been found to correct soil Zn deficiencies, although alkaline fly ash can induce Zn deficiency due to decreased solubility with increasing pH (Schnappinger *et al.*, 1975). Fly ash application has also been shown to correct B deficiencies in alfalfa (Plank and Martens, 1974). In some cases, plant yields after fly ash applications have been reduced because of B toxicity (Adriano *et al.*, 1978; Elseewi *et al.*, 1981; Zwick *et al.*, 1984; Aitken and Bell, 1985). Soil amendment with fly ash to alleviate B deficiencies should be carefully monitored in order to avoid B toxicity.

Fly ash often contains high concentrations of potentially toxic trace elements. Plants growing on soil amended with fly ash have become enriched in elements such as As, Ba, B, Mo, Se, Sr and V (Furr *et al.*, 1977; Adriano *et al.*, 1980). Although trace amounts of some of these elements are required for plant and animal nutrition, higher levels can be toxic. Highly phytotoxic elements often kill plants before they are able to accumulate large quantities of the elements, which limits their transfer to grazing animals. Elements such as Se and Mo, however, are not particularly toxic to plants and may be concentrated in plant tissue at levels that cause toxicities in grazing animals (Doran and Martens, 1972; Elseewi and Page, 1984).

Fly ash could also be an effective amendment in neutralizing soil acidity. Many of the observed chemical and biological effects of fly ash applications to soils resulted from the increased activities of Ca^{2+} and hydroxide ions. The accumulation of B, Mo, Se and soluble salts in fly ash-amended soils is apparently the most serious constraint associated with land application of fly ash (Adriano *et al.*, 1980). Fly ash application to soil not only increases pH but also increases electrical conductivity (Khan *et al.*, 1996), and cation exchange capacity (Shane *et al.*, 1988). Lime and coal fly ash application could mitigate pollution from acid deposition and improve the buffering ability of acid soils (Wu and Shao, 1996).

Fly ash application decreased maximum water holding capacity and increased availability of N, P, K and exchangeable Ca^{2+}, Mg^{2+} and trace elements such as Zn^{2+}, Cu^{2+}, Fe^{2+} and Mn^{2+}. Applications of 10 t fly ash ha^{-1} were found to be the optimal rate for improving soil properties (Kuchanwar *et al.*, 1997). Net nitrogen mineralization decreased as the rate of fly ash application increased. However, the effect of fly ash on nitrogen mineralization was also dependent on its composition. Application of 50 t ha^{-1} fly ash did not significantly affect N mineralization. Thus, high application rates of ash may be applied to soil as an alternative to disposal (Garau *et al.*, 1991).

Adsorption and desorption of phosphorus in soil amended with different rates (0-30%) of fly ash had the potential to improve plant growth by enhancing soil moisture relations (Sims et al., 1995).

Application of fly ash at 20 t ha^{-1} reduced soil crust strength from 2.38 kg cm^{-2} to 0.98 kg cm^{-2}. Application of other amendments, namely farmyard manure, sand and paddy husk either alone or in combination with fly ash also reduced the unconfined crust strength significantly (Patil et al., 1996). Fly ash addition to soils resulted in lower bulk density, reduced hydraulic conductivity and improved moisture retention at field capacity and the wilting point. Changes in soil properties might have been due to modification in macro- and micro-pore size distribution and may also have contributed to increased crop yields in light- and medium-textured soils (Kalra et al., 1998).

Fly ash, particularly when it has been collected dry and handled, contains moderate to high levels of soluble salts, primarily sulfates and borates. Dissolution of these salts into soil solution can generate high levels of salinity (> 4 m mhocm^{-1}), which can suppress plant growth or actually kill salt-sensitive seedlings and/or established vegetation. Such phytotoxicity generally decreases drastically once the ash-bound salts are leached by rainfall. The soluble salt content of an ash or ash-treated soil is measured by the electrical conductance (EC) of a water extract. Under strongly acidic conditions (< pH 5.0) ash-bound heavy metals such as Al, Mn, Zn and Cu can also come into solution and become phytotoxic (Daniels et al., 2002). Application of fly ash to soil exhibited a progressive improvement in physical properties and nutrient status, except for nitrogen levels (Singh and Siddiqui, 2002a; 2003a; Siddiqui et al., 2004). Moreover, levels of heavy metals (Pb, Ni, Cu, Mn, Co, Zn, Fe, Cr, and Cd) were significantly higher in fly ash than in soil (Siddiqui et al., 2004).

EFFECT OF FLY ASH-SOIL AMENDMENTS ON PLANTS

Seed Germination and Seedling Growth

Amendment of 10% fly ash to soil increased germination percentage of maize, sorghum, wheat and gram but germination decreased with higher ash rates except for gram, which tolerated 30% ash (Pawar et al., 1988). Seedlings of Hardwickia binata (90 days) showed better growth and were also better adapted to the growth media, in which fly ash was substituted for pond silt or farmyard manure (Mutha et al., 1997). In other findings, fly ash addition to soil delayed crop germination. This may be due to increased impedance within the soil matrix to germinating seeds. Rice and maize germination was relatively less sensitive to fly ash than that for wheat, chickpea, mustard, and lentil, and mustard, whose germination and stand establishment were strongly affected (Kalra *et al.*, 1997). Lower rates of fly ash in soil had stimulatory effects on germination in *Brassica*, whereas high concentrations caused inhibitory effects on germination. The shifts in balance of growth regulators between promoters and inhibitors may affect germination. Trace and other elements in fly ash may play important roles in different physiological processes. Increased and decreased germination was probably due to shifts in auxin levels. The increased auxin levels might be due to the presence of trace elements at approximately optimum levels and when optimum levels are exceeded, there is a shift in auxins towards the negative side of the scale, inhibiting of germination (Wong and Wong, 1989).

Plant Growth and Yield

The main constraints in the use of fly ash are its alkalinity and salt contents, which may depress plant growth and deteriorate soil quality (Hodge and Holliday, 1966). A variety of vegetables, millet and apple trees have been found to grow on potted soil amended with coal fly ash (Lisk *et al.*, 1979), because the ash is enriched in macro- and micro-nutrients which enhance growth (Planks and Martens, 1974; Martens and Beahm, 1978). Amendment of 20% fly ash with soil caused a significant increase in the number of productive tillers per hill, length of panicle, number of grains/panicle, test weight, and reduced the unfilled grains of rice. The cumulative effects of these variables ultimately led to higher grain and straw yields (Singh and Singh, 1986). Biomass production of two grasses (*Agrostis tenius* and *Festuca arundinacea*) and a legume (*Lespedeza cuneata*) was 5-30 times higher in fly ash-treated plots (Fail, 1987). Addition of poultry manure to fly ash-amended sandy soil improved soil physical and chemical properties and resulted in higher crop yields (Wong and Wong, 1987).

Amendment of soil with either fly ash or lime considerably enhanced root growth of barley and did not result in elemental toxicity to plants (Taylor, 1988). Application of 10 and 20% fly ash improved plant growth, yield and chlorophyll content of cucumber leaves but 50-100% fly ash became toxic and suppressed all measured parameters (Pasha, 1990). Soybeans grown in 25 and 50% fly ash showed significant improvement in growth, yield, leaf pigment, protein and oil contents of seeds (Singh, 1993; Singh *et al.*, 1994). Twenty to 25% fly ash and compost:soil ratio treatments generally increased growth and yield of corn. Uptake of K, Mn and Cu increased with increasing percentage (2-25%) of fly ash-compost:soil ratios. Total K content in plants was positively correlated with dry matter yield. Application of fly ash blended with compost was beneficial to corn productivity without causing any deleterious effects on growth and elemental composition (Ghuman *et al.*, 1994). Sunflower (*Helianthus annuus* L.) grown in soil treated with 1, 1.5 and

2 kg fly ash m^{-2} exhibited improved growth (Pandey et al., 1994). Relative growth rate (RGR) and net assimilation rate (NAR) increased by over 20% at low fly ash application rates and leaf area also increased. Soil amended with low rates (2%) of fly ash favoured Beta vulgaris growth and improved yields (Singh et al., 1994). Moderate doses of fly ash application (2-4% w/w) had a beneficial effect on dry matter yield of rice and also resulted in a significant increase in N, S, Ca, Na and Fe contents. However, higher doses (8% w/w) had a significant depressing effect on dry matter yields due to lower P and Zn availability and possible toxicities of other elements. The residual effect of fly ash on dry matter yield and nutrient composition of a subsequent wheat crop was not significant except for Fe content, which increased significantly from 138 mg kg^{-1} in the control to 161 mg kg^{-1} at 8% ash (Sikka and Kansal, 1995). Application of 10 t fly ash ha^{-1} was most effective in influencing yield performance of cotton, sorghum, paddy, groundnut, gram, sunflower, wheat and mustard (Matte and Kene, 1995). Tomato plants grown in the soil-ash mixture had luxuriant growth with bigger and greener leaves compared with plants grown on non-amended soil. Plant growth, yield, and chlorophyll and carotenoid contents were enhanced in 40-80% ash- amended soil; however, at 100% fly ash, yield was considerably reduced. The most economic dose of fly ash incorporation was 40%, which improved the yield and market value (mean weight) by 81 and 30%, respectively (Khan and Khan, 1996).

Fly ash incorporation with soil increased grain yield of both soybean and wheat. Percentage increase in grain yield with graded levels of fly ash (4-60%) ranged from 60 to 84 in wheat. Application of graded levels of fertilizers (50 and 100% NPK) showed similar results, especially at higher levels of ash incorporation. A considerable residual effect of fly ash was apparent on yield of wheat, but levels of ash incorporation did not vary significantly in this regard (Kumar et al., 1996). Fly ash application at 10 t ha^{-1} rendered better growth and yield of groundnut (Kuchanwar and Matte, 1997), and maximum seed yield (10.49 q ha^{-1}) of sunflower (Sugawe et al., 1997). Addition of 50% fly ash to soil increased seedling height, plant height, girth, leaf number, leaf area, spike length, and dry weight. The effect of fly ash on growth and yield are comparable with those of 10% compost or 0.6% NPK treatments (Tripathy and Sahu, 1997). Groundnut, ladies finger and radish responded positively to fly ash (15%) and yield plant^{-1} increased by 4, 57 and 77%, respectively (Sarangi and Mishra, 1998). Yield of maize, paddy and mustard increased with maximum ash addition of 10 t ha^{-1} and wheat up to an addition of 20 t ha^{-1} (Kalra et al., 1998). Fly ash, fertilizers (NPK) and farmyard manure application to soil significantly increased yield of rice but fly ash plus fertilizer proved to be more effective than fly ash + fertilizer + farmyard manure. Addition of 4% fly ash resulted in higher grain yield without any trace metal contamination of soil and plants (Kumar *et al.*, 1998). Dry weight of wheat and sunflower were highest when grown in a 1:3 fly ash : soil mixture (Malewar *et al.*, 1999).

Fly ash amendment reduced relative growth rate (RGR) and delayed wheat development (Gregroczyk, 1999). Changes in RGR mainly resulted from the variability of leaf area ratio. Rye, wheat, triticale and barley reacted variably as a result of different doses (0, 50, 100, 150 t ha^{-1}) of fly ash. Yield of rye and wheat were similar, irrespective of ash dose, and there were no distinct yield differences in triticale. Ten per cent yield increases were obtained at 100 t ha^{-1} and barley experienced a steady increase in yield with increased ash rate (Stosio and Tomaszewicz, 1999). Highest yields of grain and straw contents and uptake of nutrients, contents of crude protein and test weight were recorded with 10 t ha^{-1} fly ash applications (Bhaisara *et al.*, 2000). Tuber yield of sweet potato increased significantly with each incremental dose up to 15 t ha^{-1} fly ash (Birajdar *et al.*, 2000). Application of 20% and 40% fly ash in soil enhanced plant growth, yield, photosynthetic pigments, protein and lysine contents of wheat and rice, the maximum occurring at 40% (Singh and Siddiqui, 2002a, 2003a,c). Forty per cent fly ash amendment increased growth and yield of mustard (Siddiqui *et al.*, 2000), pea and chickpea (Singh, 2001). Fly ash application at 10 t ha^{-1} in combination with organic sources (paper factory sludge, farmyard manure and crop residue) and chemical fertilizers increased grain yield of rice, pod yield of peanut and equivalent yield of both crops, as compared to chemical fertilizer alone. There were beneficial effects of repeated ash application as compared to a one-time application at the same level, and yield advantage derived by peanut through integrated nutrient supply was greater than that for rice (Mittra *et al.*, 2003).

EFFECTS OF FLY ASH AERIAL DEPOSITION ON PLANTS

Fly ash, under humid conditions, can attach to leaves or fruit and promotes chemical as well as physical injuries. Necrotic spots appeared on leaves of many vegetative types including crops near fossil fuel-burning power plants. Vaccarino *et al.* (1983) attributed this effect to the association of vanadium with fly ash. Other studies have demonstrated that the vanadate ion acts as an inhibitor to the sodium pump (Macara, 1980) and negatively affects many enzymes (Catalan *et al.*, 1980). Fly ash particles concentrated on the surface of leaf guard cells affects the mechanism that regulates the opening and closing of the stomata and often prevents their closing by blocking the stomatal aperture (Fluckiger *et al.*, 1979; Krajickova and Mejstrick, 1984). Fly ash dusting at 2-6 g m^{-2}day^{-1} increased plant height, dry weight, chlorophyll and carotenoid contents of chickpea and wheat (Dubey *et al.*, 1982). Fly ash dusting at low rates increased plant height, metabolic rate, contents of photosynthetic pigments and dry weight of maize and soybean due to correction of boron deficiency. However, the highest dusting rates caused reductions in pigment content and dry matter production, which was attributed chiefly to excessive uptake and accumulation of boron and alkalinity caused by excessive soluble salts on the leaf surface. Moreover, a thick layer of ash also interferes with light absorption required for photosynthesis and thus reduces photosynthesis rate. Conversely, leaves laden with fly ash absorb certain wavelengths of solar radiation more effectively. Consequently, the increased temperature of the dusted leaves resulted in increased transpiration rate (Mishra and Shukla, 1986a). Concentration of photosynthetic pigments of *Datura innokia* which experienced high fly ash deposition were low in leaves, had high levels of sugars, total phenol and free proline, and high activity of oxidative enzymes (Satyanarayan *et al.*, 1988). Dusting of 2.5 and 5.0 g m^{-2}day^{-1} fly ash caused a greater increase in plant growth, yield, photosynthetic pigments, protein and lysine contents of wheat and rice, with the maximum being 5.0 g m^{-2}day^{-1} (Singh and Siddiqui, 2002b, 2003b).

ABSORPTION AND ACCUMULATION OF HEAVY METALS IN PLANTS

If fly ash has been applied to soil at sufficiently high levels, heavy metals could potentially accumulate at excessive concentrations in plant tissue. Many studies have been carried out regarding absorption and accumulation of elements in plants grown on fly-ash amended soil. Absorption of B, Cu, Co, Fe, Mg, Mn, Mo, Se and Zn by plants grown on ash-amended soils was enhanced. Concentrations of B, Ni and Se were considerably greater in eight plant species grown in ash-treated soil but uptake of Cr and Pb was unchanged (Scanlon and Duggan, 1979). Concentrations of B, Mo and Se were increased in vegetables, grains and forage crops grown on silt treated with soft coal fly ash (Gutenmann *et al.*, 1981). Application of alkaline fly ash on sludge-amended soils was effective in reducing uptake of Cd by sun grass by as much as 87%. The most apparent beneficial effect of a joint fly ash-sludge application was decreased Cd uptake from the acid soil. However, fly ash induced levels of soil salinity sufficient to damage crops. By weathering the ash before soil incorporation, the yield-depressing salinity effect was overcome (Adriano *et al.*, 1982).

Plants grown in fly ash-amended soil absorb elements present in ash, and the degree of accumulation depends upon the percentage of fly ash present in the soil (Shane *et al.*, 1988). Absorption of B, Cu, Mn and Zn by corn and soybean increased with increasing rates of fly ash application; however, P, K and Ca concentrations were not affected. The variation in soil pH induced by ash treatment could be considered the most important parameter influencing heavy metal uptake (Mishra and Shukla, 1986b). Fly ash generally decreased heavy metal concentrations in wheat seedlings. Total accumulation of heavy metals decreased with increasing ash addition in soil experiencing a higher pH increase (2 units), whereas in cases of lesser pH change (0.4 units), total metal accumulation increased. Zinc, Cr and Ni experienced low mobility in plants as evidenced by marked differences in concentrations in the aerial parts of wheat seedlings compared to roots (Petruzzelli et al., 1987). Concentrations of different trace metals, viz., Fe, Mn, Zn, Cu, Co, Pb and Ni in rice grain were highest in 8% fly ash + 10 t farmyard manure/ha + NPK. Increasing level of fly ash incorporation in soil significantly increased uptake of trace metals by rice (Kumar et al., 1998). Fly ash-treated plots had a marginally higher uptake of Zn, Cu, Fe, Mn and Cd by wheat, rice, maize and mustard (Kalra et al., 1998). Concentrations of accumulated Cu, Zn and Pb in edible parts of plants grown on ash-amended soil were within permissible limits, whereas Cd, Cr and Ni showed higher concentrations. The heterogeneous accumulation of metals in plants varies from species to species and also within the different parts of a same plant (Barman et al., 1999). The concentrations of heavy metals in shoots, roots and seeds of mustard grown in ash-amended soil were found to be within the permissible limits recommended by The Prevention of Food Adulteration Act 1995 (Siddiqui et al., 2000, 2004).

EFFECTS OF FLY ASH ON MICROORGANISMS

Due to changes in soil physico-chemical characteristics from fly ash incorporation, the activities of microorganisms may be substantially influenced. Unwheathered fly ash is essentially sterile when deposited (Cope, 1962; Rippon and Wood, 1975). Few studies of the effects on microbial populations at ash deposal sites have been conducted. Microbial numbers and diversity generally increase as ash weathers and nutrients accumulate (Rippon and Wood 1975; Klubek et al., 1992). The lower supply of organic compounds (as an energy source) and N are the principal factors which limit populations. Toxic elements present in fly ash may also adversely affect microbial activities. Several investigators have reported decreased soil respiration following ash amendments (Arthur et al., 1984; Wong and Wong, 1986; Pichtel, 1990). The reduction in microbial respiration at higher ash application rates resulted from trace element toxicity (Arthur et al., 1984). Microbial respiration decreased with increasing fly ash treatment in sandy soil, whereas, in sandy loam a depression was only recorded at higher fly ash rates (Wong and Wong, 1986). Total bacterial, actinomycete and fungal counts in the soil typically decreased with increasing ash content. Counts were depressed by 57, 80 and 86%, respectively, at a 20%

ash application rate (Pichtel and Hayes, 1990). This may be a consequence of changes in soil alkalinity, salinity or concentrations of potentially toxic elements. It may also be a result of reduced C and N contents of ash. Wheat and rice grown in fly ash-amended soil showed an increased percentage of infected leaf area by *Alternaria triticina* and *Helminthosporium oryzae* respectively (Fig. 7.1). Similarly, foliar applications of fly ash of up to 5 g plant^{-1} increased percentage infected leaf area of wheat and rice while 7.5 g fly ash application/plant deceased percentage infected leaf area (Fig. 7.2) (Singh and Siddiqui, 2002a,b; 2003a,b).

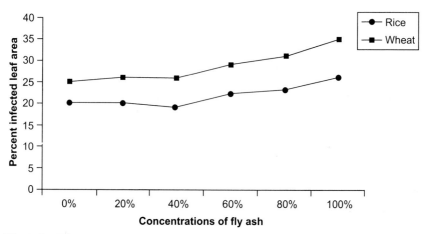

Fig. 7.1 Effect of fly ash soil amendment on leaf blight of wheat and brown spot of rice

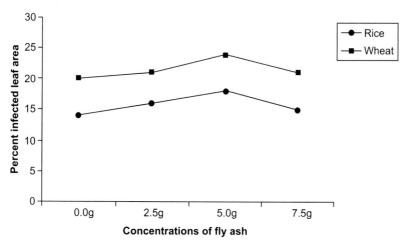

Fig. 7.2 Effect of fly ash foliar application on leaf blight of wheat and brown sport of rice

Soil phosphatase, sulphatase, dehydrogenase and invertase activities were inhibited as ash levels increased. Catalase activity was not significantly affected by ash concentration. Higher fly ash rates in soil could thus hinder normal decomposition and nutrient cycling processes. Fly ash also affects the N cycle in soil (Pichtel and Hayes, 1990). Rhizosphere microflora of *Crotolaria retusa* was indirectly affected by fly ash spraying via effects on crop physiology and root morphology, which changed root exudation patterns and ultimately affected microbial growth and activity (Panigrahi *et al.*, 1988). Soil amended with fly ash levels from 10 to 100% showed decreased root nodulation by *Rhizobium* and completely suppressed nodulation at 70 and 80% ash levels in chickpea and lentil, respectively (Singh, 1989). Plant growth, leaf chlorophyll, nitrogen and phosphorus contents, seed protein, number of VAM spore and root nodulation of chickpea and mungbean increased at low levels of fly ash (10 to 25%) and decreased with 50 to 100% rates. Root colonization by VAM fungus decreased with 10 to 100% fly ash levels. Fly ash application suppressed root colonization by VAM fungi both in single and dual inoculated plants. Spore numbers increased significantly at 20 and 40% levels, but declined from 60% onwards up to 100% ash in mycorrhiza and dual inoculated plants in comparison to their respective controls. Application of fly ash also caused significant increases in number and dry weight of nodules up to 40%, but numbers of nodules were reduced from 60% up to 100% ash in both single and dual inoculated plants (Jahan, 1993). Fly ash amendments with soil had an adverse effect on root colonization of pea by *Pseudomonas striata*. Twenty per cent fly ash caused less adverse effects than 40% ash (Siddiqui and Singh, 2005a). Amendments of 20 and 40% ash with soil also adversely affected root colonization of wheat by *Glomus mosseae*, *Pseudomonas fluorescens* and *Aspergillus awamori* (Siddiqui and Singh, 2005b).

CONCLUSIONS

Most studies suggest that there is an appropriate application range for safe utilization of fly ash for improving soil fertility and growth of various crops. The reported studies were mostly conducted under pot conditions. There is an urgent need to conduct field trials to confirm these results. The major impacts of fly ash disposal include release of toxic substances into soil and groundwater and increases of elemental concentrations in vegetation growing on ash-amended soil, and transfer of elements through the food chain. The utilization of fly ash in an integrated manner with chemical fertilizers, organic fertilizers and biofertilizers can reduce the adverse effects of fly ash on recipient soils and may interact synergistically in increasing crop growth and yield.

REFERENCES

Adriano, D.C., Page, A.L., Elseewi, A.A., Chang, A.C. (1982). Cadmium availability to sudan grass grown on soil amended with sewage sludge and fly ash. Journal of Environmental Quality 11: 197-203.

Adriano, D.C., Page, A.L., Elseewi, A.A., Chang, A.C., Straughan, I. (1980). Utilization and disposal of fly ash and other coal residues in terrestrial ecosystem : A review. Journal of Environmental Quality 9(3): 333-344.

Adriano, D.C., Woodford, T.A., Ciravolo, T.G. (1978). Growth and elemental composition of corn and bean seedlings as influenced by soil application of coal ash. Journal of Environmental Quality 7: 416-421

Aitken, R.L., Bell, L.C. (1985). Plant uptake and phytotoxicity of boron in Australian fly ashes. Plant Soil, 84: 245-257.

Arthur, M.A., Zwick, T.C., Tolle, D.A., van Voris, P. (1984). Effects of fly ash on microbial CO_2 evolution from an agricultural soil. Water Air and Soil Pollution 22: 209-216.

Barman, S.C., Kisku, G.C., Bhargava, S.K. (1999). Accumulation of heavy metals in vegetables, pulse and wheat grown in fly ash amended soil. Journal of Environmental Biology 20(1): 15-18.

Bhaisare, B., Matte, D.B., Badole, W.P., Deshmukh, A. (2000). Effect of fly ash on yield, uptake of nutrients and quality of green gram grown on vertisol. Journal of Soils and Crops 10(1): 122-124.

Birajdar, R.R., Chalwade, P.B., Badole, S.B., Hangarge, D.S., Shelage, B.S. (2000). Effect of fly ash and FYM on nutrient availability soil and yield of sweet potato. Journal of Soils and Crops 10(2): 248-251.

Cary, E.E., Gilbert, M., Bache, C.A., Gutenmann, W.H., Lisk, D.J. (1983). Elemental composition of potted vegetables and millet grown on hard coal bottom ash-amended soil. Bulletin of Environmental Contamination and Toxicology 31: 418-423.

Catalan, R.E., Martinez, A.M., Aragones, M.D. (1980). Effects of vanadate on the cyclic AMP-protein kinase system in rat liver. Biochemistry Biophysics Research Communication 96: 673-678.

Chang, A.C., Lund, L.J., Page, A.L., Warneke, J.E. (1977). Physical properties of fly ash-amended soils. Journal of Environmental Quality 6(3): 267-270.

Cope, F. (1962). The development of a soil from an industrial waste ash. In: Soil Science and Society. pp. 859-863. Trans. Comm. IV, V. Int. Soc. Soil Sci., Palmerstown, New Zealand.

Daniels, W.L., Stewart, B., Haering, K., Zipper, C. (2002). The potential for beneficial reuse of coal fly ash in Southwest Virginia Mining environments. *Reclamation Guidelines*, Powell River Project. Virginia Cooperative Extension Publication. 460.

Doran, J.W., Martens, D.C. (1972). Molybdenum availability as influenced by application of fly ash to soil. Journal of Environmental Quality 1: 186-189.

Dubey, P.S., Pawar, K., Shringi, S.K., Trivedi, L. (1982). Effects of fly ash deposition on photosynthetic pigments and dry matter production of wheat and gram. Agro-Ecosystem 8(2): 137-140.

Elseewi, A.A., Page, A.L. (1984). Molybdenum enrichment of plants grown on fly ash - treated soils. Journal of Environmental Quality 13: 394-398.

Elseewi, A.A., Bingham, F.T., Page, A.L. (1978). Availability of sulfur in fly ash to plants. Journal of Environmental Quality 7: 69-73.

Elseewi, A.A., Grimm, S.R., Page, A.L., Straughan, I.R. (1981). Boron enrichment of plants and soils treated with coal ash. Journal of Plant Nutrition, 3, 409-427.

Elseewi, A.A., Straughan, I.R., Page, A.L. (1980). Sequential cropping of fly ash - amended soils: Effects on soil chemical properties and yield and elemental composition of plants. Science of Total Environment 15: 247-259.

Fail, J.L., Jr. (1987). Growth response of two grasses and a legume on coal fly ash-amended strip mine spoils. Plant Soil 101: 149-150.

Fluckiger, W., Oertli, J.J., Fluckiger-Keller, H. (1979). Relationship between stomatal diffusive resistance and various applied particle sizes on leaf surfaces. Z. Pflanzenphysiology 91: 173-175.

Fulekar, M.H., Naik, D.S., Dave, J.M. (1983). Heavy metals in Indian coals and corresponding fly ash and their relationship with particle size. International Journal of Environmental Studies 21(2): 179-182.

Furr, A.K., Parkinson, T.F., Hinrichs, R.A., Van Campben, D.R., Bache, C.A., Gutenmann, W.H., St. John, Jr. L.E., Pakkala, I.S., Usk, D.J. (1977). National survey of elements and radioactivity in fly ashes: Absorption of elements by cabbage grown in fly ash- soil mixtures. Environmental Science Technology 11: 1194-1201.

Garau, M.A., Dalmau, J.L., Felipo, M.T. (1991). Nitrogen mineralizaiton in soil amended with sewage sludge and fly ash. Biology and Fertility of Soils., 12(3): 199-201.

Ghuman, G.S., Menon, M.P., Chandra, K., James, J., Adriano, D.C., Sajwan, K.S. (1994). Uptake of multi-elements by corn from fly ash-compost amended soil. Water Air and Soil Pollution 72: 285-295.

Gregorczyk, A. (1999). Growth analysis of spring wheat growing in the soil substrate at different content of fly ash. Roczniki Nauk Rolniczych Seria A, Produkcja Roslinna., 114(12): 13-24.

Gutenmann, W.H., Elfving, D.C., Valentino, D.I., Lisk, D.J. (1981).Trace element absorption on soil amended with soft-coal fly ash. Bio Cycle 22.

Hansen, L.D., Fisher, G.L. (1981). Elemental distribution in coal fly ash particles. ES and T., 14(9): 1111-1117.

Helder, T., Stulterheim, E., Olte, K. (1982). The toxicity and toxic potential of fly ash from municipal incinerators assessed by means of fish life stage test. Chemosphere 11: 968-972.

Hermann, J., Wisniewski, W. (1988). Mobile aluminium in sour soils fertilized with power plant fly ashes. Environmental Protection Engineering 14(2): 85-90.

Hill, M.J., Lamp, C.A. (1980). Use of pulverized fuel ash from Victorian brown coal as a source of nutrients for a pasture species. Australian Journal of Experimental Agriculture and Animal Husbandary 20: 377-384.

Hodgson, D.R., Holliday, R. (1966). The agronomic properties of pulverized fuel ash. Chem. Ind. (May) pp. 785-790.

Jahan, K. (1993). Effects of some air pollutants on root-nodulation and VAM colonization of roots of some leguminous crops. Ph.D thesis, Aligarh Muslim University, Aligarh.

Jones, C.C., Amos, D.F. (1976). Physical changes in Virginia soils resulting from additions of high rates of fly ash. In: Proc. 4th Int. Ash Utilization Symposium. Eds. J.H. Faber, A.W. Babcock and J.D. Spencer, U.S. Energy Research Development Administration MERC/SP-76-4. Morgantown.

Kabata-Pendias, A., Adriano, D.C. (1995). In: Soil Amendments and Environmental Quality, Eds J.E. Rechcigl, Lewis Publichers, Boca Raton FL.

Kalra, N., Jain, M.C., Joshi, H.C., Choudhary, R., Herat, R.C., Vasta, B.K., Sharma, S.K., Kumar, V. (1998). Fly ash as a soil conditioner and fertilizer. Bioresource Technology 64: 163-167.

Kalra, N., Joshi, H.C., Chaudhary, A., Choudhary, R., Sharma, S.K. (1997). Impact of fly ash incorporation in soil on germination of crops. Bioresource Technology 61: 39-41.

Khan, M.R. and Khan, M.W. (1996). The effect of fly ash on plant growth and yield of tomato. Environmental Pollution 92: 105-111.

Khan, S., Tahira, B., Singh, J. (1996). Effect of fly ash on physico-chemical properties and nutrient status of soil. Indian Journal of Environmental Health 38: 41-46.

Kim, A.G., Kazonich, G. (2003). Mass release of trace elements from coal combustion by-products. Fly Ash Library Home: http://www.fly ash.info.

Klubek, B., Carlson, C.L., Oliver, J., Adriano, D.C. (1992). Characterization of microbial abundance and activity from three coal ash basins. Soil Biologyand Biochemistry 24: 1119-1125.

Krajickova, A., Mejstrik, V. (1984). The effect of fly ash particles on the plugging of stomata. Environmental Pollution 36(1): 83-93.

Kuchanwar, O.D., Matte, D.B. (1997). Study of graded doses of fly ash and fertilizers on growth and yield of groundnut Arachis hypogaea. Journal of Soils and Crops 7(1): 36-38.

Kuchanwar, O.D., Matte, D.B., Kene, D.R. (1997). Effect fly ash application on physico-chemical properties of soil. Journal of Soils and Crops 7(1): 73-75.

Kumar, A., Sarkar, A.K., Singh, R.P., Sharma, V.N. (1996). Effect of fly ash and fertilizer levels on yield and trace metal uptake by soybean and wheat crops. Journal of Indian Society of Soil Science 47(4): 744-748.

Kumar, A., Sarkar, A.K., Singh, R.P., Sharma, V.N. (1998). Yield and trace metal levels in rice (Oryza sativa) as influenced by fly ash, fertilizer and farmyard manure application. Indian Journal of Agriculture Science 68(9): 590-592.

Lisk, D.J., Furr, A.K., Parkinson, T.F. (1979). Elemental content of apple, millet and vegetables grown in pots of neutral soil amended with fly ash. Journal of Agriculture and Food Chemistry 27(1): 135-138.

Macara, I.G. (1980). Vanadium - an element in search of a role. Trends in Biochemical Science April pp. 92-94.

Malewar, G.U., Adsul, P.B., Ismail, S. (1999). Effect of different combinations of fly ash and soil on growth attributes of forest and dryland fruit crops. Indian Journal of Forest 21(2): 124-127.

Martens, D.C., Beahm, B.R. (1976). Growth of plants in fly ash amended soils. p. 657-664. In: Proc. 4th Int. Ash Utilization Symposium. Eds.J.H. Faber et al., St. Louis, MO 24-25 Mar. 1976. MERC SP-76/4. ERDA Morgantown Energy Res. Center, Morgan-town.

Martens, D.C., Beahm, B.R. (1978). Chemical effects on plant growth of fly ash incorporation into soil. In: Environmental Chemistry and cycling processes. Eds.D.C. Adriano and I.I. Brisin. ERDA Symp. Sea. Conf. 7604229. U.S. Department of Commerce, Spring and Faild, VA.

Martens, D.C., Schnappinger, Jr., M.G., Zelazny, L.W. (1970). The plant availability of pot assium in fly ash. Proceeding of American Society of Soil Science 34: 453-456.

Matte, D.B., Kene, D.R. (1995). Effect of fly ash application on yield performance of kharif and rabi crops. Journal of Soils and Crops 5(2): 133-136.

Mattigod, S.V., Dhanpat R., Eary, L.E., Ainsworth, C.C. (1990). Geochemical factors controlling the mobilization of inorganic constituents from fossil fuel combustion residues: I. Review of the major elements. Journal of Environmental Quality 19: 188-201.

Mengel, K., Kirkby, E.A. (1982). Principles of Plant Nutrition, International Potash Institute, Bern, Switzerland.

Mishra, L.C., Shukla, K.N. (1986a). Effects of fly ash deposition on growth, metabolism and dry matter production of maize and soybean. Environmental Pollution 42: 1-13.

Mishra, L.C. and Shukla, K.N. (1986b). Elemental composition of corn and soybean grown on fly ash amended soil. Environmental Pollution 12: 313-321.

Mittra, B.N., Karmakar, S., Swain, D.K., Ghosh, B.C. (2003). Fly Ash – a potential source of soil amendment and a component of integrated plant nutrient supply system. Int. Ash Utilization Symposium, Center of Applied Energy Research, University of Kentucky, Paper # 28. http://www. fly ash. info.

Moliner, A.M., Street, J.J. (1982). Effect of fly ash and lime on growth and composition of corn (*Zea mays* L.) on acid sandy soils. Proceeding of Soil Crop Science Society of Florida 41: 217-220.

Mutha, N., Burman, U., Kumar, P., Aggarwal, R.K., Harsh, L.N. (1997). Effect of fly ash substitution in nursery mixture on growth of *Harwickia binata*. Indian Journal of Forestry 20 (3): 213-219.

Onliveros, J.L., Clapp, T.L., Kosson, D.S. (1990). Physical properties and chemical species distribution within municipal waste combustor ashes. Environmental Progress 8(3): 200-206.

Page, A.L., Elseewi, A.A., Straughan, I. (1979). Physical and chemical properties of fly ash from coal-fired power plants with reference to environmental impacts. Residue Review 71: 83-120

Pandey, V., Misra, J., Singh, S.N., Singh, N., Yunus, M., Ahmad, K.J. (1994). Growth response of *Helianthus annus* L. grown on fly ash amended soil. Indian Journal of Environmental Biology 15: 117-125.

Panigrahi, A., Mukherji, K.G., Mc Michael, B.L., Persson, H. (1988). Microbial ecology of the rhizosphere microflora of *Crotolaria retusa* L. in relation to fly ash. Plant roots and their environment: Proceeding of An ISRR – Symposium, August 21-26, Uppsala, Sweden, 314-318.

Pasha, M.J. (1990). Studies on interaction of some air pollutants, *Sphaerotheca fuliginea* and *Meloidogyne javanica* on cucumber. Ph.D. Thesis, Aligarh Muslim University, Aligarh, India.

Patil, C.V., Math, K.K., Bulbule, A.V., Prakash, S.S., Yeledhalli, N.A. (1996). Effects of fly ash on soil crust strength and crop yield. Journal of Maharashtra Agricultural University 21(1): 9-11.

Pawar, K., Dubey, P.S., Verma, S.P. (1988). Germination behaviour of some important crop species in fly ash incorporated soils. Advancement of crops and monitoring of environment. Progress in Ecology 10: 295-305.

Petruzzelli, G., Lubrano, L., Cervelli, S. (1987). Heavy metal uptake by wheat seedlings grown in fly ash amended soils. Water Air and Soil Pollution 32: 389-395.

Pichtel, J.R. (1990). Microbial respiration in fly ash/sewage sludge-amended soils. Environmental Pollution 63: 225-237.

Pichtel, J.R., Hayes, J.M. (1990). Influence of fly ash on soil microbial activity and populations. Journal of Environmental Quality 19: 593-597.

Plank, C.O., Martens, D.C. (1974). Boron availability as influenced by application of fly ash to soil. Proceeding of American Society of Soil Science 38: 974-977.

Plank, C.O., Martens, D.C., Hallock, D.L. (1975). Effect of soil application of fly ash on chemical composition and yield of corn (*Zea mays* L.) and on chemical composition of displaced soil solutions. Plant Soil 42: 465-476.

Rippon, J.E., Wood, M.J. (1975). Microbiological aspects of pulverized fuel ash. In: The Ecology of Resource Degradation and Renewal. pp. 331-349. Eds. M.J. Chadwick and G.T. Goodman. John Wiley and Sons, New York.

Roy, W.R. Thiery, R.G., Schuller, R.M., Suloway, J.J. (1981). Coal Fly Ash: A review of the literature and proposed classification system with emphasis on environmental impacts. Environmental Geology Notes 96. Illinois State Geological Survey. Champaign.

Sarangi, P.K., Mishra, P.C. (1998). Soil metabolic activities and yield in groundnut, ladies finger and radish in fly ash amended soil. Research Journal of Chemistry and Environment 2(2): 7-13.

Satyanarayana, G., Bhatnagar, A.R., Acharya, V.H. (1988). Effect of fly ash pollution on *Datura innokia*. Environmental Ecology 6: 92-95.

Sawell, S.E., Bridle, T.R., Constable, T.W. (1989). Heavy metal leachability from solid waste incinerator ashes. Waste Management Research 6(3): 227-238.

Sawyer, T., Bandiera, S., Safe, S., Hutzinger, D., Olie, K. (1983). Bioanalysis of polychlorinated dibenzofurans and dibenzo-p-dioxins mixture in fly ash. Chemosphere 12: 529-536.

Scanlon, D.H., Duggan, J.C. (1979). Growth and element uptake of woody plants on fly ash. Environment Science and Technology 13: 311-315.

Schnappinger, Jr., M.G., Martens, D.C., Plank, C.O. (1975). Zinc availability as influenced by application of fly ash to soil. Environment Science and Technology 9: 258-261.

Schramm, K.W., Wu, W.Z., Henkelmann,B., Merk, M., Xu,Y., Zhang, Y., Kettrup, A.(1995). Influence of linear alkaylbenzene sulfonate (LAS) as organic co-solvent on leaching behaviour of PCDD/Fs from fly ash and soil. Chemosphere 31 (6):3445-3453.

Sear, L.K.A., Weatherley, A.J., Dawson, A. (2003). The environmental impact of using fly ash– the UK producers' perspective. Int. Ash Utilization Symposium, Center for Applied Energy Research, University of Kentucky, Paper # 20. http:/www.flyash.info.

Sengupta, P. (2002). Fly ash for acidic soils. Science and Technology. *The Hindu*. Online edition of India's National Newspaper. Feb. 28th.

Shane, B.S., Littman, C.B., Essick, L.A., Gutenmann, W.H., Doss, G.J., Lisk, D.J. (1988). Uptake of selenium and mutagens by vegetables grown in fly ash containing greenhouse media. Journal of Agriculture and Food Chemistry 36: 328-333.

Siddiqui, S., Ahmad, A., Hayat, S. (2004). The fly ash influence the heavy metal status of the soil and the seeds of sunflower-A case study. Journal of Environmental Biology 25(1): 59-63.

Siddiqui, S., Singh, L.P., Khan, M.W. (2000). Influence of fly ash on growth and yield of mustard plants (*Brassica juncea* Czern & Coss). Chemical and Environmental Research 9: 303-308.

Siddiqui, Z.A., Singh, L.P. (2005a). Effects of fly ash, *Pseudomonas striata* and *Rhizobium* sp. On the reproduction of *Meloidogyne incognita* and on the growth and transpiration of pea. Journal of Environmental Biology 26: 117-122.

Siddiqui, Z.A., Singh, L.P. (2005b). Effects of fly ash and soil microorganisms on the plant growth, photosynthetic pigments and leaf blight of wheat. Journal of Plant Disease and Protection 112: 146-155.

Sikka, R., Kansal, B.D. (1995). Effect of fly ash application on yield and nutrient composition of rice, wheat and on pH and available nutrient status of soils. Bioresource Technology 51: 199-203.

Sikka, R., Kansal, B.D., Beri, V., Chaudhary, M.R., Sidhu, P.S., Pashricha, N.S., Bajwa, M.S. (1994). Potential use of fly ash as a source of plant nutrient. Proeeding of International Symposium of Nutrition and Management Sustained Production 2: 136-137.

Sims, J.T., Vasilas, B.L., Ghodrati, M. (1995). Evaluation of fly ash as a soil amendment for the Atlantic Costal Plain: II. Soil chemical properties and crop growth. Water Air and Soil Pollution 81(3-4): 363-372.

Singh, K. (1993). Impact assessment of root-knot nematode on air pollution stressed plants. Ph.D. Thesis, Aligarh Muslim University, Aligarh, India.

Singh, L.P. (2001). Effect of fly ash on growth, yield and root nodulation of pea and chickpea. Bulletin of Pure and Applied Science 20: 45-48.

Singh, L.P., Siddiqui, Z.A. (2002a). Effects of fly ash and *Alternaria triticina* on the yield, protein and lysine contents of three cultivars of wheat. Thai Journal of Agricultural Science 35: 397-405.

Singh, L.P., Siddiqui, Z.A. (2002b). Effects of foliar deposition of fly ash and *Helminthosporium oryzea* on growth, yield, photosynthetic pigments, protein and lysine contents of three cultivars of rice. Journal of Environmental Studies and Policy 5: 41-51.

Singh, L.P., Siddiqui, Z.A. (2003a). Effects of fly ash and *Helminthosporium oryzea* on growth and yield of three cultivars of rice. Bioresource Technology 87: 73-78.

Singh, L.P., Siddiqui, Z.A. (2003b). Effects of *Alternaria triticina* and foliar fly ash deposition on growth, yield, photosynthetic pigments, protein and lysine contents of three cultivars of wheat. Bioresource Technology 86: 189-192.

Singh, L.P., Siddiqui, Z.A. (2003c). Effects of fly ash and *Helminthosporium oryzea* on yield, photosynthetic pigments, protein and lysine contents of three cultivars of rice. Proceeding of 12th National Symposium on Environment June 2003, H.N. Bahuguna Garhwal University, Teri Garhwal, India, pp. 272-279.

Singh, N., Singh, S.N., Yunus, M., Ahmad, K.J. (1994). Growth response and element accumulation in *Beta vulgaris* L. raised in fly ash-amended soils. Ecotoxicology 3: 287-298.

Singh, N.B., Singh, M. (1986). Effect of balanced fertilization and fly ash levels on the yield and yield attributes of rice under saline condition. Fertilizer News 47-48.

Singh, S.K. (1989). Studies on interaction of air pollution and root-knot nematodes on some pulse crops. Ph.D. Thesis, Oligarchs Muslim University, Aligarh, India.

Sovocool, G.W., Mitchum, R.K., Tondeur, Y., Munslow, W.D., Vonnahme, T.L., Donnelly, J.R. (1988). Bromo- and bromochloro-polynuclear aromatic hydrocarbons, dioxins and dibenzofurans in municipal incinerator fly ash. Biomedical Environment Mass Spectrum 15: 669-676.

Srivastava, A. (2002). The fly ash burden. Industry & Technology. Down to Earth. July 15th, pp. 23-24.

Srivastava, V.K., Srivastava, P.K., Kumar, R., Misra, U.K. (1986). Seasonal variations of metals in coal fly ash. Environmental Pollution 11(2): 83-89.

Stosio, M., Tomaszewicz, T. (1999). Impact of addition of coal ash from a power plant "Dona Odra" on chemical properties of medium soil and yield of winter crops. X Miedzynarodowa Konferencja Naukowa "Aktualne problemy inzynierii rolniczej", Miedzyzdroje, Polska, 17 19 Czerwca 1998 r. Tom I. Inzynieria Rolnicza.: 257-262.

Sugawe, G.T., Quadri, S.J., Dhoble, M.V. (1997). Repsonse of sunflower to the graded levels of fertilizers and fly ash. Journal of Maharashtra Agricultural University 22: 318-319.

Sumner, M.E., Fey, M.V., Noble, A.D. (1991). In: Soil Acidity, (Eds B. Ulrich and M.E. Sumner), Springer-Verlag, New York.

Taylor, Jr., E.M., Schuman, G.E. (1988). Fly ash and lime amendment of acidic coal spoil to aid revegetation. Journal of Environmental Quality 17: 120-124.

Theis, T.L., Wirth, J.L. (1977). Sorptive behavior of trace metals on fly ash in aqueous systems. Environment Science and Technology 11: 1096-1100.

Tolle, D.A., Arthur, M.F., Pomeroy, S.E. (1982). Fly Ash Use for Agriculture and Land Reclamation: A critical literature review and identification of additional research needs. RP-1224-5. Battelle Columbus Laboratories. Columbus.

Tripathy, A., Sahu, R.K. (1997). Effect of coal fly ash on growth and yield of wheat. Journal of Environmental Biology 18(2): 131-135.

Vaccarino, C., Cimino, G., Tripodo, M.M., Langana, G., Giudice, L.L., Matarese, R. (1983). Leaf and fruit necroses associated with vanadium rich ash emitted from a power plant burning fossil fuel. Agriculture Ecosystem and Environment 10: 275-283.

Van Hook, R.I. (1979). Potential health and environmental effects of trace elements and radionuclides from increased coal utilization. Environmental Health Perspective 33: 227-247.

Wadge, A., Hutton, M. (1987). The leachability and chemical speciation of selected trace elements in fly ash from coal combustion and refuse incinerators. Environmental Pollution 48: 85-99.

Wallace, A., Wallace, G.A. (1986). Enhancement of the effect of coal fly ash by a poly acryl amide soil conditioner on growth of wheat. Soil Science 141: 387-389.

Wallace, A., Alexander, G.V., Soufi, S.M., Mueller, R.T. (1980). Micronutrient supplying power of pyrite and fly ash. Journal of Plant Nutrition 2: 147-153.

Wong, J.W.C., Wong, M.H. (1987). Co-recycling of fly ash and poultry manure in nutrient deficient sandy soil. Resource Conservation 13: 291-304.

Wong, M.H., Wong, J.W.C. (1986). Effects of fly ash on soil microbial activity. Environmental Pollution 40: 127-144.

Wong, M.H., Wong, J.W.C. (1989). Germination and seedling growth of vegetable crops in fly ash amended soils. Agriculture Ecosystem and Environment 26: 23-35.

Wu, E.J., Choi, W.W. and Chen, K.Y. (1982). Chemical affiliation of trace metals in coal ash. "Emission Contr. from Stationary Power Sources: Tech., Econ. and Environ. Assessments" Published by AM Institute of Chemical Engineering 76 (201): 177-187.

Wu, J.M., Shao, F. (1996). Effects of alkaline ameliorator additions on the acidic buffering abilities of aluminium rich Alfisols. Jornal of Environmental Science 8(2): 235-241.

Zwick, T.C., Arthur, M.F., Tolle, D.A., van Voris, D. (1984). A unique laboratory method for evaluating agro-ecosystem effects of an industrial waste product. Plant Soil 77: 395-399.

CHAPTER 8

Effects of Metal-Contaminated Organic Wastes on Microbial Biomass and Activities: A Review

K. Chakrabarti[1], P. Bhattacharyya[2] and A. Chakraborty[3]

[1]Institute of Agricultural Science, Calcutta University, 35, Ballygunge Circular Road, Kolkata – 700019, West Bengal, INDIA.
[2]Dept. of Geology & Geophysics, Indian Institute of Technology, Kharagpur-721302, West Bengal, INDIA.
[3]Department of Agronomy, Bidhan Chandra Krishi Viswavidyalaya, P.O. Mohanpur, Dist: Nadia-741242, West Bengal, INDIA.

Soil quality is not defined solely by its physical and chemical parameters, but is also intimately linked to microbial parameters. Several microbial parameters have been examined to study the effects of metal toxicity due to recycling of metal contaminated-municipal wastes like sewage sludges and municipal solid wastes, in arable soils. Major emphasis has been placed on effects of sewage sludge. Microbial parameters studied include microbial biomass C, soil respiration, fluorescein diacetate hydrolyzing activity and non-symbiotic N_2-fixation (all serving as estimates of microbial activity). Most studies are from temperate climates; unfortunately, little information is available under tropical conditions.

Metal-contaminated waste applications, in most cases, result in high soil metal loadings, which disrupt microbial biomass as well as its activities. However, in some cases, there are no ill effects or even positive effects from such applications. The data seems to reflect metal effects as a function of soil type, metal type and metal forms. The ratio of microbial biomass C to soil organic C and biomass-specific respiration are useful indices in metal toxicity evaluations, both under controlled field experiments as well as in studies where there is no real control.

INTRODUCTION

Enhancement or maintenance of soil quality is essential for sustainable agriculture and for a safe environment for future generations. Soil quality does not depend solely on physical and chemical properties of soil, but is intimately related to microbiological parameters (Elliot et al., 1996). In soil, microbial biomass has special significance. Biomass serves as the source and sink of plant nutrients in soil. Biomass makes a critical contribution to nutrient flows, organic matter turnover and structural stability of soil aggregates. Due to its dynamic nature, microbial biomass responds quickly to soil management and perturbations (Carter, 1986) and to the local soil environment (Skopp et al., 1990; Duxbury and Nkambule, 1994).

Pollution of soil by heavy metals through the addition of metal containing organic wastes or mining or smelter emission is a real threat to soil quality. Heavy metals once entering into soil persist over long period of time. Although microorganisms require metals for growth and activity, they are toxic to the microorganisms when they are present in excessive concentrations. Thus, heavy metal toxicity to microorganisms in soil has been studied in great extent and reported. Present review deals with the effect of metal contaminated organic wastes on soil microbial biomass and its activities as determined by soil respiration, fluorescein diacetate hydrolyzing activity and non-symbiotic nitrogen fixation.

MICROBIAL BIOMASS IN METAL-CONTAMINATED SOIL

Development of methods which readily estimate soil microbial biomass has greatly facilitated the study of the impact of heavy metals on soil microorganisms. Microbial biomass is the component of soil organic matter that constitutes living microorganisms smaller than $5-10\,\mu m^3$. Microorganisms include bacteria, actinomycetes, fungi, alga, protozoa and microfauna but not plant roots and fauna larger than $5-10\,\mu m^3$ (Sparling, 1985). Microbial biomass data are expressed usually as biomass C in $\mu g/g$ oven dry soil. Data on microbial biomass provides an early warning of soil pollution by toxicants, heavy metals, pesticides etc. Several studies carried out both under laboratory (Dar, 1996; Khan and Scullion, 1999; Rost et al., 2001) as well as field conditions (Brookes, 1995) revealed the detrimental influences of heavy metals on soil microorganisms. Bååth (1989) and Giller et al. (1998), however, opined that laboratory studies bear little relation to most "real life" contamination of soils. Metal concentrations in soil are gradually built up over a long period of time due to atmospheric metal deposition, the use of fertilizers or pesticides, or land application of contaminated wastes. In laboratory studies, an unrealistic dose is often applied at a single time, thus confounding the results.

Brookes and McGrath (1984) used the biomass concept to investigate the residual effects of heavy metals from past applications (1942 – 1967) of sewage sludge to a loamy sand at Woburn Experimental Farm on microbial biomass and organic matter dynamics. Controlled field experiments revealed that the amount of microbial biomass in soil, whether receiving sewage sludge or sludge containing composts, was smaller than in soils which received farmyard manure over the same period. The effect was attributed to the presence of toxic metals (at approximately the current European Union permitted limit) in the sludges and was readily detectable more than 20 years after the final sludge application. In soils low in metal content there was no detectable change in microbial biomass.

The effect of heavy metals on microbial biomass was investigated in 30 urban soils from Japan, contaminated mainly with Zn and Pb to different extents. Microbial biomass was not significantly affected by the readily soluble Zn + Pb concentrations. Microbial carbon was positively related to cation exchange capacity, total organic carbon, total nitrogen and the numbers of fungal colonies present (Ohya et al., 1988). Schuller (1989) found no significant effect of heavy metals on microbial biomass in an old landfill site that received wastes from a copper smelter, and in an urban beach forest.

Chander and Brookes (1991; 1993) based on field experiments in the United Kingdom with metal-rich sludges and metal-free sludge reported that both Cu and Zn concentrations at about 2 – 3 times the current European Union permitted limits decreased soil microbial biomass. Copper and Zn decreased biomass by about 40% and 30 – 40% respectively, compared to soils receiving metal-free sludge.

Perucci (1992) observed that the addition to field soil of metal-rich municipal solid waste compost at 30 and 90 t ha^{-1} for three years resulted in a significant increase in microbial biomass C. The increase was much higher at higher rates of application.

The effect of roadside metal contamination on the size of soil microbial biomass was examined from five sites in London. There was no evidence that increased soil concentrations of Pb, Cu, Ni, Zn and Cd close to the road adversely affected microbial biomass, and the larger biomass at a 0.5 m distance from the roadside was attributed to a positive response to higher soil pH and carbon levels at this distance (Post and Beeby, 1993).

Long-term field experiments (1980 – 1990) with sewage sludge in Germany revealed that low metal sludge imparted beneficial effects on microbial biomass, while higher heavy metal contamination of soils substantially decreased biomass (Flieâbach et al., 1994). Aoyama and Nagumo (1996) observed that microbial biomass C expressed on a soil organic carbon basis was negatively correlated with the amount of 0.1 M HCl-extractable Cu from apple orchard soil contaminated with Cu, Pb and As. Banerjee et al. (1997) examined the effect of metal rich sewage sludge (0, 50 and 100 t ha^{-1} on dry matter basis) on microbial biomass C in a clay loam soil in Canada. After six years of sludge application, either in single or repeated applications, there was a general increase in microbial biomass C. Similar observations were also noted by Pascual et al. (1999).

Chander *et al.* (2001) analyzed soils of lower Saxony, Germany, variously contaminated with sewage sludge, sediment and mining and factory residues to assess the effects of different heavy metal sources and different metal species on microbial biomass C. The biomass C did not necessarily decrease with increasing heavy metal content, reflecting the importance of other environmental factors, e.g. differences in carbon input.

Bhattacharyya (2002) studied the effect of addition of metal-rich Calcutta municipal solid waste compost on microbial biomass C in waterlogged rice soil from 1997 – 1999. Municipal solid waste compost imparted a positive effect on soil microbial biomass. The author recommended comprehensive field trials before thorough conclusions could be drawn on the use of municipal solid waste compost in agriculture.

PROPORTION OF MICROBIAL BIOMASS C IN SOIL ORGANIC C IN METAL-POLLUTED SOIL

Microbial biomass C is a component of soil organic C. Generally, microbial biomass C comprises about 1 – 4% of soil organic C (Jenkinson and Ladd, 1981; Anderson and Domsch, 1989). Limited controlled field experiments (Chander *et al.*, 1993; Flieâbach *et al.*, 1994) revealed the usefulness of this proportional value to detect metal stress on microbial biomass. However, such studies are required to be extended to soils which have been subjected to 'actual' environmental pollution, and under a range of climatic situations. In such situations the link between microbial biomass C to total organic C can itself constitute an internal control. When soils deviate significantly from these ratio values, which are perceived as typical for the particular climatic condition, it indicates that some damage to the normal functioning of the ecosystem has occurred (Brookes, 1995; Giller *et al.*, 1998).

MICROBIAL ACTIVITY IN METAL-CONTAMINATED SOIL

Soil Respiration

Soil respiration is one of oldest methods (and still frequently used) for quantifying microbial activity in soil. Soil respiration is measured in the absence of added substrate (basal respiration) or in the presence of added substrate (substrate-induced respiration). Both techniques have been extensively used as a measure of metal toxicity to microbial activity under field as well as laboratory studies (Giller *et al.*, 1998).

Both increases (Leita *et al.*, 1995; Saviozzi *et al.*, 1997; Pascual *et al.*, 1999; Khan and Scullion, 1999) and decreases (Doelman and Haanstra, 1984;

Ohya *et al.*, 1988; Hattori, 1992; Post and Beeby, 1996) in soil respiration have been measured in soils treated with metals. Murray *et al.* (2000) found that microbial activity, as measured by substrate-induced soil respiration was not affected by metal contamination in the Montreal area of Canada.

Biomass-Specific Respiration

The ratio of basal soil respiration to microbial biomass carbon is termed as the specific respiration of the biomass or qCO_2 (Anderson and Domsch, 1985). This parameter reflects the physiological characteristic of the microbial community in its environment (Anderson, 1994). Brookes (2000) emphasizes that non-experimental field data is difficult to interpret due to lack of suitable "controls". Under such cases biomass-specific respiration measurements may constitute an internal control and provide an early warning of soil disturbances.While increases in the ratio as a result of metal addition has been measured in certain studies (Brookes and McGrath, 1994; Flieâbach *et al.*, 1994; Bardgett and Saggar, 1994; Leita *et al.*, 1995; Aoyama and Nagumo, 1997) decreases in other studies (Valsecchi *et al.*, 1995; Insam *et al.*, 1996; Chander *et al.*, 2001) have also been recorded. The increase in the ratio values has been explained as that metal causes a diversion of energy form biosynthesis to microbial activity in the soil microbial biomass. The decrease in specific respiration rate in metal polluted soil compared to non-polluted soil could be due to (1) microorganisms in metal polluted and non-polluted soil use different substrates, (Insam *et al.*, 1996), (2) large proportion of older microorganisms in dormant phase or a change in community structure from bacteria to fungi, in metal polluted soil (Chander *et al.*, 2001).

The stated contradictory results indicate that the biomass specific respiration as an indicator of metal stress on microorganisms is not straightforward. The results should be interpreted cautiously. There is a need in the refinement of methodology and experimental approach.

Fluorescein Diacetate Hydrolyzing Activity

Measurement of soil enzyme activities has been often used as a measure of microbial activity. However, Nannipieri *et al.* (1990) are of the view that the measure of a single enzyme's activity may not represent overall soil microbial activity, because enzymes are substrate-specific and soil microbial populations may not release enzymes at equal rates. To overcome this problem, hydrolysis of fluorescein diacetate has been advocated as a promising method of determining soil microbial activity (Dick, 1994). Such hydrolysis is not specific to a specific enzyme, but is mediated by a number of enzymes like proteases, lipases and esterases, at a single time. This assay has not been widely used in the study of metal stress on microorganisms in soil, but has sufficient merit for adoption.

Generally, enzyme activities are positively correlated with soil organic carbon (Gianfreda and Bolllag, 1996). Dick (1994) suggested that a correct comparison between soils under different environmental conditions and management could be made on the basis of enzyme activities to organic carbon status of soil. Aoyama and Nagumo (1996) also suggested that enzyme activity to soil organic matter basis could become useful indicators for assessing the effects of heavy metals in soil differing in organic matter content.

Perucci (1992) studied the rate of fluorescein diacetate hydrolysis in a loamy soil amended with metal-rich solid municipal refuse over a three year period under field conditions. Addition of higher rates of the compost caused a significant increase in activity. Mitra (2004) observed that contaminated landfill soils of Kolkata, India, had higher fluorescein diacetate hydrolyzing activity than the nearby uncontaminated soil, because of higher amount of organic carbon in the landfill soils. The ratio of fluorescein diacetate hydrolyzing activity to soil organic carbon in the landfill soils was lower than the uncontaminated soil, indicating the metal stress on enzymes.

Non-symbiotic Nitrogen Fixation in Metal-polluted Soil

In soil, biological nitrogen fixation has special significance as it complements chemical fertilization, thereby reducing environmental pollution due to fertilizer use. There is certain information to suggest that measurement of nitrogen fixation could be a suitable test for soil pollution by heavy metals.

Soils of Kirghizia in the USSR showed a decrease in free-living nitrogen fixation activity, which was correlated with increasing amounts of soil Pb and Zn (Leutonova et al., 1985). Skujins et al. (1986) found nitrogen fixation to be extremely sensitive to pollution in laboratory experiments using Cu and Cr. Brookes et al. (1986) observed that the addition of metal-contaminated sewage to soil over a period of twenty years resulted in severe reduction in the rate of heterotrophic nitrogen fixation, as judged by acetylene reduction and N^{15}- labeled N_2 tests. Mårtenson and Witter (1990) studied the effects of addition of various organic materials including sewage sludge and nitrogen fertilizers on leguminous bacteria, blue green algal populations and free-living N_2 fixing soil bacterial populations in a 30 year old field experiment. Sewage treated soils experienced reduced nitrogen fixing activity compared to its control and this reduction was attributed to metal buildup from sewage application.

In Sweden, Dahlin et al. (1997) observed that sewage sludge application for 20 years increased concentrations of soil Cd, Cu, Pb and Zn but had not attained the current European community limits for soils. Reductions of 15 and 8% in autotrophic and heterotrophic acetylene reduction potential were observed. Concentrations of heavy metals in soil close to or below current UK permissible limits led to 94% inhibition of heterotrophic and 98% reduction of cyanobacterial N_2 fixation activity (Lorenz *et al.*, 1992). Assessment of heterotrophic N_2 fixation was found to be more promising in this respect than measurement of cyanobacterial activity, as results were obtained more quickly and easily.

CONCLUSIONS

Several investigations have been conducted regarding the effects of metals in soil from aerial deposition, sewage sludge and municipal solid waste applications on soil microbial biomass and activities. In response to the addition of metal-rich organic wastes to soils, increases as well as decreases in microbial parameters are evident. This point to the fact that metal toxicity to microbial biomass and its activities are dependent on metal forms and their concentrations in the environment. Permissible limits for metal loading to soil in temperate climates have been fixed from the point of view of plant and animal health but not with regards to microbial activity and health. Furthermore, most information in this respect is available only from temperate climatic regions. In tropical countries, including India where municipal wastes are generated in abundance, research on proper utilization of wastes in crop production without compromising soil health/quality, is needed and should be undertaken in different agroclimatic regions.

REFERENCES

Anderson, T.H. (1994). Physiological analysis of microbial communities in soil: applications and limitations. In: Beyond the biomass. Compositional and functional analysis of soil microbial communities. Ritz, K., Dighton, J. and Giller, K.E. (eds.). pp. 67-76. John Wiley, Chichester.

Anderson, T.H., Domsch, K.H. (1985). Determination of ecophysiological maintenance requirements of soil microorganisms in a dormant state. Biology and Fertility of Soils. 1: 81-89.

Anderson, T.H., Domsch, K.H. (1989). Ratios of microbial biomass carbon to total organic carbon in arable soils. Soil Biology and Biochemistry 21: 471-479.

Aoyama, M., Nagumo, T. (1996). Factors affecting microbial biomass and dehydrogenase activity in apple orchard soils with heavy metal accumulation. Soil Science and Plant Nutrition 42: 821-831.

Aoyama, M., Nagumo, T. (1997). Comparison of the effects of Cu, Pb and As on plant residue decomposition, microbial biomass and soil respiration. Soil Science and Plant Nutrition 43: 613-622.

Bååth, E. (1989). Effects of heavy metals in soil on microbial processes and populations. Water, Air and Soil Pollution 47: 335-379.

Banerjee M.R., Burton, D.L., Depoe, S. (1997). Impact of sewage sludge application on soil biological characteristics. Agriculture Ecosystem Environment 66: 241-249.

Bardgett, R.D., Saggar, S. (1994). Effects of heavy metal contamination on the short-term decomposition of labeled [^{14}C] glucose in a pasture soil. Soil Biology and Biochemistry. 26: 727-733.

Bhattacharyya, P. (2002). Evaluation of Calcutta municipal solid waste compost for crop production. Ph.D. Thesis, University of Calcutta, India.

Brookes, P.C. (1995). The use of microbial parameters in monitoring soil pollution by heavy metals. Biology and Fertility of Soils 19: 269-279.

Brookes P.C., McGrath S.P. (1994). Effects of metal toxicity on the size of the soil microbial biomass. Journal of Soil Science 35: 341-346.

Brookes, P.C., McGrath, S.P., Heijnen, C. (1986). Metal residues in soils previously contaminated with sewage sludge and their effects on growth and nitrogen fixation by blue-green algae. Soil Biology and Biochemistry 18: 345-353.

Brookes, P.C. (2000). Changes in soil microbial properties as indicators of adverse effects of heavy metals. Memorie di Scienzi Fisiche e Naturali. 24: 205-227.

Carter, M.R. (1986). Microbial biomass as an index for tillage induced changes in soil biological properties. Soil Tillage Research 7: 29-40.

Chander, K., Brookes, P.C. (1991). Effects of heavy metals from past application of sewage sludge on microbial biomass and organic matter accumulation in a sandy loam and silty clay loam U.K. soil. Soil Biology and Biochemistry 10: 927-932.

Chander, K., Brookes, P.C. (1993). Effects of Zn, Cu and Ni in sewage sludge on microbial biomass in a sandy loam soil. Soil Biology and Biochemistry 25: 1231-1239.

Chander, K., Dyckmans, J., Joergensen, R.G, Meyer, B., Raubuch, M. (2001). Different sources of heavy metals and their long term effects on soil microbial properties. Biology and Fertility of Soils. 34: 241-247.

Dahlin, S., Witter, E., Martenson, A., Turner, A., Bååth, E. (1997). Whrer's the limit? Changes in the microbiological properties of agricultural soils at low levels of metal contamination. Soil Biology and Biochemistry 29: 1405-1415.

Dar, G.H. (1996). Effects of cadmium and sewage sludge on soil microbial biomass and enzyme activities. Bioresource Technology 56: 209-218.

Dick, R.P. (1994). Soil enzyme activities as indicators of soil quality. In: Defining Soil quality for sustainable environment. Special Pub. 35. Eds. J.W. Doran, D.C. Coleman, D.F. Bezdicek, and B.A. Stewart, Soil Science Society of America. Inc., Madison.

Doelman, P., Haanstra, L. (1984). Short-term and long-term effects of cadmium, chromium, copper, nickel, lead and zinc on soil microbial respiration in relation to abiotic soil factors. Plant Soil 79: 317-327.

Duggan, J.C., Wiles, C.C. (1976). Effects of municipal composts and nitrogen fertilizer on selected soils and plants. Compost Science 17: 24-31.

Duxbury, J.M., Nkambule, S.V. (1994). Assessment and significance of biologically active soil organic nitrogen. In: Defining soil quality for suatainable environment. pp. 126-146. Eds. J.W. Doran, D.C. Coleman, D.F. Bezdicek, and B.A. Stewart, Soil Science Society of America. Inc., Madison.

Elliot, L.F., Lynch J.M., Papendick, R.I. (1996). The microbial component of soil quality. In: Soil Biochemistry. pp. 1-21 G. Stotzky, and J.M. Bollag, Marcel Dekker, Inc. New York.

Flieâbach, A., Martens, R., Reber, H.H. (1994). Soil microbial biomass and microbial activity in soils treated with heavy metal contaminated sewage sludge. Soil Biology and Biochemistry 26: 1201-1205.

Gianfreda, L., Bollag, J.M. (1996). Influence of metal and anthropogenic factors on enzyme activity in soil. In: Soil Biochemistry. pp. 123-193. G. Stotzky, and J.M. Bollag, Marcel Dekker, Inc. New York.

Giller, K.E., Witter, E., McGrath, S.P. (1998). Toxicity of heavy metals to microorganisms and microbial processes in agricultural soils: a review. Soil Biology and Biochemistry 30: 1389-1414.

Hattori, H. (1992). Influence of heavy metals on soil microbial activities. Soil Science and Plant Nutrition 38: 93-100.

Insam, H., Parkinson, D., Domsch, K.H. (1989). Influence of microclimate on soil microbial biomass. Soil Biology and Biochemistry 21: 211-221.

Insam, H., Hutchin, T.C., Reber, H.H. (1996). Effects of heavy metal stress on the metabolic quotient of the soil microflora. Soil Biology and Biochemistry 28: 691-694.

Jenkinson, D.S., Ladd, J.N. (1981). Microbial biomass in soil: measurement and turnover. In: Soil Biochemistry. pp. 415-417. Eds. E.A. Paul, and J.N. Ladd. New York.

Khan, M., Scullion, J. (1999). Microbial activity in grassland soil amended with sewage sludge containing varying rates and combinations of Cu, Ni and Zn. Biology and Fertility of Soils. 30: 202-209.

Leita, L., De Nobili, M., Muhlbachova, G., Mondini, C., Marchiol, L., Zerbi, G. (1995). Bioavailability and effect of heavy metals on soil microbial biomass survival during laboratory incubation. Biology and Fertility of Soils. 19: 103-108.

Leutonva, S.V., Umarov, M.M., Niyazova, G.A., Mebkhin, Y.I. (1985). Nitrogen fixation activity as a possible criterion for determining permissible concentration of heavy metals in soils. Soviet Soil Science 17: 88-92.

Lorenz, S.E., McGrath, S.P., Giller, K.E. (1992). Assessment of free-living nitrogen activity as a biological indicator of heavy metal toxicity in soil. Soil Biology and Biochemistry 24: 601-606.

Martenson, A.M., Witter, E. (1990). Influence of various soil amendments on nitrogen-fixing soil organisms in a long-term field experiment, with special reference to sewage sludge. Soil Biology and Biochemistry 22: 977-982.

Mitra, A. (2004). Studies on microbiological and related aspects of landfill soil. Ph.D. Thesis, University of Calcutta, India.

Murray, P., Ge, Y., Hendershot, W.H. (2000). Evaluating three trace element contamination sites: a field and laboratory investigation. Environmental Pollution 107: 127-135.

Nannipieri, P., Gregos, S., Ceccanti, B. (1990). Ecological significance of the biological activity in soil. In: Soil Biochemistry. pp 293-354. Eds. J.L. Smith and E.A. Paul. Marcel Dekker, Inc., New York.

Ohya, H., Fujiwara, S., Komai, Y., Yamaguchi, M. (1988). Microbial biomass and activity in urban soils contaminated with Zn and Pb. Biology and Fertility of Soils 6: 9-13.

Pascual, J.A., Garcia, C., Hernandez, T. (1999). Lasting microbiological and biochemical effects of the addition of municipal solid waste to an arid soil. Biology and Fertility of Soils 30: 1-6.

Perucci, P. (1992). Enzyme activity and microbial biomass in a field soil amended with municipal refuse. Biology and Fertility of Soils 14: 54-60.

Post, R.D. and Beeby, A.N. (1993). Microbial biomass in suburban roadside soils: estimated based on extracted microbial C and ATP. Soil Biology and Biochemistry 25: 199-204.

Post, R.D. and Beeby, A.N. (1996). Activity and microbial decomposer community in metal contaminated roadside soils. Journal of Applied Ecology. 33: 703-709.

Rost, U., Joergensen, R.G., Chander, K. (2001). Effects of Zn enriched sewage sludge on microbial activities and biomass in soil. Soil Biology and Biochemistry 33: 633-638.

Saviozzi, A., Levi-Minzi, R., Cardelli, R., Riffaldi, R. (1997). The influence of heavy metals on carbon dioxide evolution from a typic xerochrept soil. Water, Air and Soil Pollution 93: 409-417.

Schuller, E. (1989). Enzyme activities and microbial biomass in old landfill soils with long-term metal pollution. Verhandlungen – Gesellschaftfur – Okologie. 18: 339-348.

Skopp, J., Lawson, M.D., Doran, J.W. (1990). Steady state aerobic microbial activity as afunction of soil water content. Soil Science Society American. Journal 54: 1619-1625.

Skujins, J., Nohrstedt, H., Oden, S. (1986). Development of a sensitive biological method for development of a low level toxic contamination in soils. 1. Selection of nitrogenase activity. Swedish Journal of Agricultural Research 16: 113-118.

Sparling, G.P. (1985). The Soil Biomass. In : Soil Organic matter and biological activity. pp. 223 Eds. D.Vaughan, and R.E.Malcolm. Martinus Nijoff / Dr. W. Junk, Dordrecht. Boston, Lanchester.

Valsecchi, G., Gigliotti, C. and Farini, A. (1995). Microbial biomass, activity and organic matter accumulation in soils contaminated with heavy metals. Biology and Fertility of Soils 20: 253-259.

Werner, W. (1996). Biotransfer of heavy metals as a function of site-specific and crop-specific factors. Plant Research and Development. 43: 31-49.

CHAPTER 9

Characterization and Evaluation of Municipal Solid Waste Compost by Microbiological and Biochemical Parameters in Soil under Laboratory and Field Conditions

P. Bhattacharyya[1], K. Chakrabarti[2] and A. Chakraborty[3]

[1]Dept. of Geology & Geophysics, Indian Institute of Technology, Kharagpur-721302, West Bengal, INDIA.
[2]Institute of Agricultural Science, Calcutta University, 35, Ballygunge Circular Road, Kolkata – 700019, West Bengal, INDIA.
[3]Department of Agronomy, Bidhan Chandra Krishi Viswavidyalaya, P.O. Mohanpur, Dist: Nadia-741242, West Bengal, INDIA.

Municipal solid wastes (MSW) are generally considered as environmental nuisances, including being a heavy source of heavy metal pollution. However, MSW contain substantial amounts of biodegradable matter and plant nutrient elements. Composting MSW is an attractive method of waste disposal and resource recovery. We studied the suitability of Kolkata municipal solid waste compost (MSWC) application to soil with the perspective of metal pollution hazards on microbial biomass-C, soil respiration and some enzyme activities at two water regimes under laboratory and field conditions.

Composts, derived from Kolkata municipal solid wastes, were mature as revealed by data on $C/N_{organic}$ of water extracts, NH_4^+ and NO_3^- ion contents, CEC, humus-C, humification ratio and humification index. Low respiration and microbial biomass-C as percentage of compost organic-C reflected the stability of the product. The product did not contain indicator organisms like *E.coli* and *Salmonella* spp. The compost contained substantial amounts of total Zn, Cu, Pb and Cd. DTPA-and water-extractable Zn, Cu and Pb were present appreciably but Cd was below detection limits.

The MSWC, compared against traditionally used cow dung manure (CDM), and control at 60% water holding capacity (WHC) of soil and waterlogged condition over a period of 120 days using Gangetic alluvial soil (Typic Fluvaquent), increased microbial biomass-C (MBC), substrate induced soil respiration (SR), urease activity (UA) and acid phosphatase activity (PA) with the increase in graded dose from 2.5 to 40 t ha^{-1} at both water regimes compared against the control. The CDM (20 and 40 t ha^{-1}) recorded higher values of these parameters than did MSWC at the corresponding application rates. At 60% WHC soil, MBC and SR of the treated soils increased up to 30 days of incubation (DOI) and then gradually declined but maintained higher values than the control soil until the end of the incubation. The activities of both soil enzymes increased up to 60 DOI and then gradually declined. Under waterlogged conditions MBC, UA and PA of the amended soils decreased during the entire period of incubation, but still maintained higher values than the control soil. The SR of soils, irrespective of treatments, progressively increased up to 120 DOI.

The field experiments were carried out during the rainy seasons of 1997, 1998 and 1999 with rice (IET- 1444) under waterlogged conditions in Gangetic alluvial soil, with treatments: control; MSWC (60 kg N ha^{-1}); CDM (60 kg N ha^{-1}); MSWC (30 kg N ha^{-1}) + U (urea, 30 kg N ha^{-1}); CDM (30 kg N ha^{-1}) + U (30 kg N ha^{-1}) and F (60:30:30 NPK ha^{-1}). The MBC, MBC as percentage of organic-C, UA, and PA activities were higher in CDM than MSWC-treated soils, due to a higher amount of biogenic organic materials like water-soluble organic carbon, carbohydrates and mineralizable nitrogen in the former.

The appreciable amounts of heavy metals in MSWC indicated no evidence of any detrimental influence on microbiological and related soil parameters in this short-term study. However, long-term comprehensive field experiments are necessary before thorough conclusions can be made.

INTRODUCTION

Tropical soils are generally low in organic matter content (Kanwar, 1976). Organic matter is an important attribute of soil quality. Prudent action is warranted in maintaining soil organic matter status. In the face of a global decline in organic content of soil and paucity of traditional organic matter supplements, MSWC is gaining popularities as an alternate organic soil supplement. This should also provide for the effective disposal of solid waste, since, through the recycling of a potential resource for soil amelioration, we would be a step closer towards sustainable agriculture (Parr and Hornick, 1992). The ultimate goal of composting is to produce a stable, mature humus-like product that can be successfully used for soil improvement and crop growth. While stability is a function of biological activity, maturity indicates the presence or lack of phytotoxic organic acids (Epstein, 1997). Improper maturity and stability of these composts (Jime'nez and Garcia, 1989) and their heavy metals and pathogens content (Stratton et al., 1995) deter their widespread use. Composition of MSWC varies widely, depending on the quality of the waste generated. Both beneficial and adverse effects of MSWC have been catalogued (Gallardo-Lara and Nogales, 1987; Pal and Bhattacharyya, 2003). A major objection to the use of MSWC has been due to its potentially hazardous heavy metals contents (Mitra *et al.*, 2003). These may impair soil quality indicators like microbial biomass (Brookes *et al.*, 1986; Ghosh *et al.*, 2004) and their activities (Tyler, 1981; Roy *et al.*, 2004). Although some studies have been conducted on the effect of MSWC on these aspects, little information is available under waterlogged conditions, such as is encountered in rice paddies. Waterlogging often alters chemical and microbial processes that influence nutrient cycling and accumulation of toxins. A reduced environment may cause more metals to go into soil solution (Bhattacharyya *et al.*, 2003a; Ghosh and Bhattacharyya, 2004), which may inhibit enzyme activities, and this become toxic to microorganisms (Pulford and Tabatabai, 1988; Gianfreda and Bollag, 1996). This chapter describes (a) the characteristics of Kolkata MSWC in respect of stability and maturity, and (b) the effect of addition of MSWC to soil, compared to CDM and no input, on soil MBC, SR, UA and PA under laboratory and field conditions.

CHARACTERIZATION OF THE MSWC

The samples were dark brown and smelled like forest soil, deemed necessary for mature composts (Epstein, 1997). Water-holding capacity of the samples was very high (88-97%) due to the presence of humified carbon and clay (27-33%). Reaction (pH) of the samples varied from 7.3 to 7.4, which indicated maturity (Jime'nez and Garcia, 1989). The high electrical conductivity of the samples (2.7 to 2.8 dS m^{-1}) may raise soil salinity level on long-term application. The samples contained organic-C in the range of 102 to 118 g kg^{-1}, which was lower than the average value (300 g kg^{-1}) in Europe and the US (He *et al.*, 1992). Mean total-N content of the samples (9 g kg^{-1}) was within the range of 4.5 to 16.7 g kg^{-1} in Germany and Hong Kong respectively (He *et al.*, 1992). C/N ratio (8-12) and C/N-organic water extract (5.6-6.3) of the samples were consistent to the proposed standard of maturity (Chanyasak *et al.*, 1982; Chakrabarti *et al.*, 2003). The limit value for ammonium < 0.04% (Zucconi and Bartoldi, 1987) and nitrate > 300 mg kg^{-1} (Forster *et al.*, 1993) were met by the samples. The total P and K contents of the samples were optimum. The cation exchange capacity of the samples exceeded the limit value of 60 cmol (p^+)/kg on an ash free-basis (Harada and Inoko, 1980). Humification indices are a good measure of the state of degradation and maturity of compost. Mean values of humus-C/organic-C (26%), humic acid-C/organic-C (17.6%) and humic acid-C/fulvic acid-C (2.11) indicated that the samples were sufficiently humified and mature

(Saviozzi et al., 1988; Bhattacharyya et al., 2004). The samples contained high total and bioavailable (DTPA-and water-extractable) heavy metals (Table 9.2).

Stability of the compost sample indicates the change in decomposition of organic matter and is a function of biological activity. Mean respiration rate and the microbial biomass-C as a percentage of organic-C of the samples (Table 9.1) were much lower than the limit value (5 mg CO_2-C/g compost-C/ 4 days) proposed by Epstein (1997) and 1.7% by Mondini et al. (1997), respectively, for compost maturity stability. The samples contained high microbial populations as well as enzyme activities (Table 9.1). The absence of E. coli and Salmonella spp. in the samples indicated that the generation of high temperatures during the early phase of composting killed the harmful microorganisms.

MICROBIAL BIOMASS-C

Laboratory Study

The soils receiving organic (i.e., CDM and MSWC) treatments showed significantly higher levels of MBC than the control soil (Figures 9.1 and 9.2). CDM-treated soils showed significantly higher levels of MBC than did the MSWC-treated samples. There was also a corresponding significant increase in MBC of soil with the increase in dose from of 2.5 to 40 t/ha. The trend of periodic variation in soil MBC with water regimes at 60% WHC (WH) and waterlogged condition (WL) varied, except with the no input treatment, wherein a continuous decrease was obtained from 15 days of incubation. During the incubation in WH regime, the MBC in soil increased steadily up to 30 days, where as afterwards, it declined steadily to its lowest level by day 120. In contrast, under the WL regime, the decrease in peak values was observed from 15 days onwards. The mean value of MBC in soil was statistically significant in WH than WL regime.

Significant variation between the mean MBC values at different periods was also observed (P < 0.05). The highest and lowest values were at 15 and 120 days of incubation respectively.

Higher MBC in the CDM enriched soils over the MSWC-treated soils testifies to the qualitative difference between the two materials (Sakamoto and Oba, 1991; Bhattacharyya et al., 2001). Water soluble carbon, carbohydrate and mineralizable N were higher in CDM (Table 9.1). These acted as energy sources for the soil microorganisms, hence contributing to the increase in biomass. Nutrients in soil are essential for microbial proliferation. However, shortage of nutrients in the control soils at both WH and WL regimes created a periodic decreasing trend of MBC. In addition to nutrient shortage, microorganisms in WL regime experienced oxygen depletion stress and

Table 9.1 Characteristics of municipal solid waste compost

Parameter	MSWC	CDM	Soil
pH	7.35–7.40	6.11	5.5
EC (dS/m)	2.70–2.78	2.17	0.294
Sand (%)	38–44	–	25.64
Silt (%)	26–30	–	28.64
Clay (%)	27–30	–	45.22
Organic Carbon (g/kg)	102–118	110	13.9
Total-N (g/kg)	8–11	9.5	1.7
C/N	11–12	11.6	8.2
C/N-organic of water extract	5.6–6.3	–	–
Water soluble carbon (mg/kg)	542	679	–
Carbohydrate content (μg glucose/100 g)	600	1300	–
Mineralizable nitrogen (mg/kg)	160	230	–
Ammonium nitrogen (mg/kg)	58–96	–	–
Nitrate nitrogen (mg/kg)	1180–1650	–	–
Total phosphorous (mg/kg)	4000–5000	4920	60
Available phosphorous (mg/kg)	261–304	352.3	6.2
Total potassium (g/kg)	16–19		
Available potassium (g/kg)	2.13–2.5	2.37	0.107
C.E.C (cmol (p$^+$)/kg)	71–106	68.4^z	19^x
Humus-C/Organic-C (%)	24–28	–	–
Humic acid-C/Organic-C (%)	17–19	–	–
Humic acid-C/Fulvic acid-C	1.78–2.64	–	–
Respiration(μg CO_2-C/gcompost-C/d at 22°C)	220–320	–	–
Microbial biomass-C	1.35–1.62	–	–
Organic-C (%)	(1.53±0.5)	–	–
Bacteria ($\times 10^8$/g)	16–114	–	–
Fungi ($\times 10^3$/g)	21–61	–	–
Actinomycetes ($\times 10^4$/g)	17–49	–	–
Azotobacter ($\times 10^5$/g)	33–102	–	–
E.coli	(–)ve	–	–
Salmonella	(–)ve	–	–
Urease(μg urea hydrolysed g^{-1} compost h^{-1} at 37°C)	126-279	–	–
Phosphatase (μg pnp released g^{-1} compost h^{-1} at 37°C)			
Acid	612–880	–	–
Alkaline	1009–1999	–	–

Table 9.2 Heavy metals concentration of municipal solid waste compost

Parameters	MSWC	CDM	Soil
Zn (mg kg^{-1})			
Total	487-599	170	50
DTPA extractable	94.27-233	4	16
Water extractable	4-9.5	B.D.L	0.05
Cu (mg kg^{-1})			
Total	101-190	10	21
DTPA extractable	36-74	0.25	4
Water extractable	7.5-19.5	B.D.L	0.03
Pb (mg kg^{-1})			
Total	214-319	28.99	22
DTPA extractable	87-167	0.12	3
Water extractable	11-19	B.D.L	0.05
Cd (mg kg^{-1})			
Total	1-3	B.D.L	0.75
DTPA extractable	-	B.D.L	B.D.L
Water extractable	-	B.D.L	B.D.L

B.D.L. Below detection limit

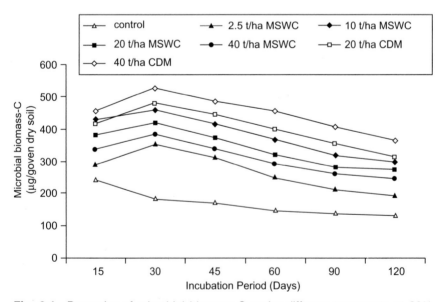

Fig. 9.1 Dynamics of microbial biomass-C under different treatments at 60% WHC of soil

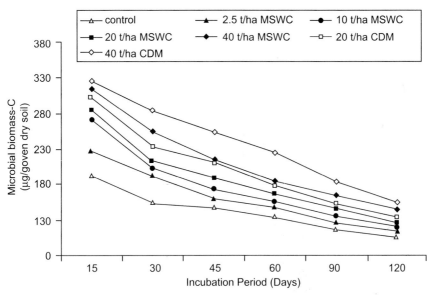

Fig. 9.2 Dynamics of microbial biomass-C in soil under different treatments under the waterlogged condition

other detrimental influences (Ponnamperuma, 1972). This caused further reduction in soil MBC. Under a waterlogged condition, the soil environment can be virtually dominated by the anaerobic or facultative microorganisms that usually have minimal cell yield (Tate, 2000).

A gradual increase followed by a decrease in soil MBC, treated with CDM or MSWC up to 30 days at WH regime may have been related to the availability of biogenic materials for biomass stimulation (Jenkinson and Ladd, 1981). Another reason may be the incorporation of exogenous microorganisms (Perucci, 1992). Decrease in soil MBC after 30 days of incubation can be associated with nutrient shortage or "protective capacity" of soil for biomass (Sparling, 1985). At 120 days of incubation organic matter treated soils had higher MBC than did the soil sample with no input. This could be the consequence of partial protection of biomass by the humic substances in CDM or MSWC (Pascual et al., 1997). The soil MBC decreased from 15 days of incubation under the WL condition irrespective of treatment. This may have originated largely from the anaerobic bacteria with oxygen depletion. Nutrient shortage was probably not a limiting factor as the anaerobic bacteria are less efficient in carbon assimilation (Alexander, 1977). Higher MBC levels, at all stages of incubation, in soil enriched with MSWC than in soils with no input indicate an improvement in the microbiological component of soil. Apparently there was no of interference by the heavy metals present in the MSWC.

Field Study

The results indicated statistically significant ($p < 0.05$) differences were measured in the level of MBC in the soil between various treatments and periods in the study years (Figure 9.3). The pattern of variation of MBC in the soil during the three years of study was similar. The significantly highest soil MBC was obtained with CDM + Urea followed by CDM > MSWC + Urea > MSWC> Fertilizer > Control. Integrated application of urea with CDM or MSWC increased the soil MBC substantially compared to separate applications. Application of CDM increased the soil MBC than MSWC did, when applied alone, or in combination with urea. The soil MBC periodically declined until 60 DAT, corresponding to the reproductive stage of rice. After the rice was harvested, the soil MBC tended to increase. The significantly highest soil MBC was detected at 15 DAT, irrespective of treatments.

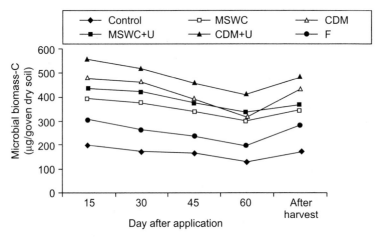

Fig. 9.3   Temporal variation of microbial biomass-C in rice soil under different treatments in 1997

The determination of MBC by the fumigation-extraction method (Vance *et al.*, 1987) is a reliable measure for soils that are acidic, recently amended with organics and submerged (Joergensen, 1995). All these conditions prevailed in this study. Furthermore, MBC is a more responsive measure than soil organic carbon. The addition of organics, like CDM or MSWC, significantly ($P<0.05$) increased soil MBC. Soil systems receiving more organic matter tend to harbor higher levels of MBC with greater microbial activity (Sparling, 1985). The higher level of MBC found in fertilizer treated soils compared to the control treatment, might have resulted from greater root mass and exudates from the crop, which stimulated the proliferation of microorganisms.

Microbial biomass-C was higher in CDM than MSWC-treated soils. As stated earlier, it was not the quantity of organic matter but the quality with respect to decomposition, which was important for proliferation of microorganisms in soil. In this study, almost the same quantity of organic-C was added to soil through MSWC or CDM. Higher levels of MBC in CDM-treated soil could be due to greater amounts of biogenic materials in CDM than MSWC. These biogenic materials include mineralizable N, water soluble C and carbohydrate (Table 9.1).

Leita *et al.* (1999) and Bhattacharyya *et al.* (2003b) found similar results with integrated application of organics and fertilizers. Fertilizers may meet the demand of mineral nutrition by microbes but cannot provide carbon, which is a major constituent of microbial cells. Integrated application of organic and inorganic materials provides balanced mineral nutrients as well as carbon. This is reflected by increased levels of soil MBC in CDM+Urea and MSWC+Urea plots.

Witt *et al.* (2000) noted a similar temporal pattern of soil MBC, as observed herein, due to treatment variation. They noted a gradual decrease in soil MBC, irrespective of treatments, from 15 DAT up to 60 DAT. This decline may reflect competition between growing crop roots and microorganisms for nutrients (Bhattacharyya *et al.*, 2003b). It is also possible that the trend may reflect the detrimental influences of submergence on microorganisms (Bhattacharyya *et al.*, 2003c). In submerged conditions the plant effect on soil MBC is small. Laboratory incubation of soil under waterlogged condition also resulted in decreasing trend of MBC. Observed increase in MBC of soil, irrespective of treatments, after crop harvest could be explained as the microorganisms did not have to face the stress of waterlogging. In addition, the left over root mass of the crop was at the disposal of the biomass. It became the new substrate for the proliferation of microorganisms. This is reflected in higher MBC of soil after crop harvest.

MICROBIAL BIOMASS-C AS PERCENTAGE OF ORGANIC-C (RATIO INDEX VALUE, RIV)

Field Study

The RIV increased from 1997 to 1999 in all the treatments (Table 9.3). The lowest RIV was noted in the control plots (1.21 in 1997 and 1.44 in 1999), while the highest occurred in CDM+Urea plots (3.36 in 1997 and 3.63 in 1999). The MSWC treatment either alone, or with urea, recorded higher RIV than the fertilizer treatment and control.

The RIV range of 1.21 to 3.63 in different treatments was comparable to previous reports (Anderson and Domsch, 1980). The application of

Table 9.3　Variation in microbial biomass-C (MBC) as percentage of organic-C in rice soil under different treatments (average of 4 replicates) after crop harvest during 1997 to 1999

Treatment	Organic C (g/kg)			MBC (μg/g)			MBC/organic C (%)		
	1997	1998	1999	1997	1998	1999	1997	1998	1999
Control	14	13.8	13.5	169	190	194	1.19	1.22	1.23
MSWC	14.5	15	15.4	342	390	405	2.35	2.38	2.44
CDM	14.6	15.2	15.7	431	473	476	2.83	2.95	3.11
MSWC+U	14.2	14.7	15.1	366	442	457	2.34	2.41	2.62
CDM+U	14.3	14.8	15.3	480	540	556	2.95	3.21	3.31
F	14.27	13.9	13.6	276	280	296	1.96	2.11	2.12

exogenous organics (CDM or MSWC) elevated the RIV. Those values were further increased by integrated application along with urea. This indicated accumulation of organic matter and improvement in nutrient status of soil, as microbial biomass is a labile reservoir of plant nutrients (Jenkinson and Ladd, 1981).

SOIL RESPIRATION

Laboratory Study

Soils enriched with CDM or MSWC, at all the doses, showed increased SR as compared to the control soil (Figures 9.4 and 9.5). Soil application of CDM significantly increased SR compared with MSWC, both under waterlogged and 60% WHC of soil. The periodic variation of SR was not similar in WL and WH regimes, excepting the no input treatment During 30 days of incubation the SR values reached a peak and subsequently decreased to the lowest values at 120 days in the WH regime.

Incorporation of carbon substrate through CDM or MSWC instantaneously increased SR. High adenylate charge (a measure of metabolic energy stored in the adenosine nucleotide pool – Brookes, 1995), maintained by the organisms, partly illustrate their ability to respond readily to exogenously added substrates. The growth and activity are concomitant to the decline in substrate availability. The small amount of substrate is used for maintenance (Sparling, 1985). The same phenomenon was observed in the WH regime; however, a gradual increasing trend of SR under the WL regime in all treatments probably reflected the carbon assimilating potential of the microorganisms (Alexander, 1977). Under the WL situation, the anaerobic bacteria dominated the soil systems. These are poor assimilators of organic

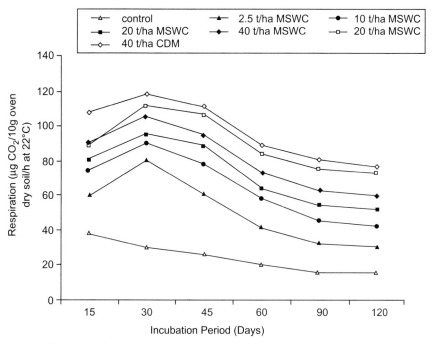

Fig. 9.4 Soil respiration under different treatments at 60% WHC

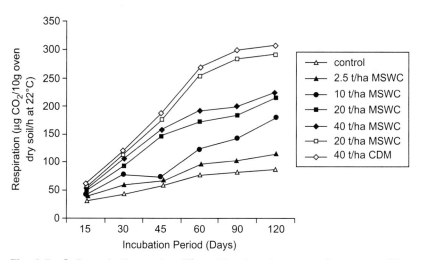

Fig. 9.5 Soil respiration under different treatments at waterlogged condition

carbon, evolving more waste products like CO_2 and other low molecular weight organic compounds. Higher SR in CDM or MSWC compared to input treatment pertains to the higher MBC in the supplemented soils.

SOIL ENZYME ACTIVITIES

Laboratory Study

Significantly increased urease and acid phosphatase activities were observed in soil samples supplemented with CDM or MSWC (Figures 9.6, 9.7, 9.8 and 9.9). The CDM-supplemented soil samples showing higher activities than

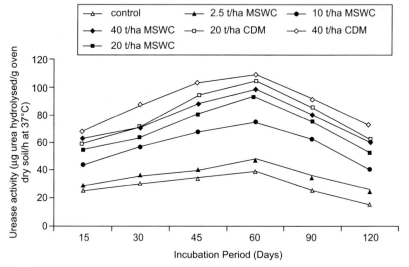

Fig. 9.6   Urease activity under different treatments at 60% WHC of soil

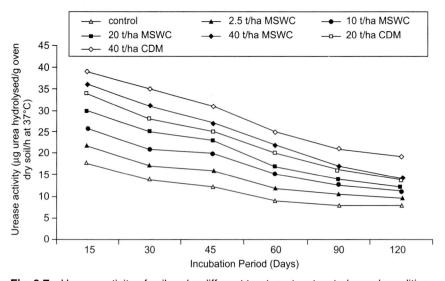

Fig. 9.7   Urease activity of soil under different treatments at waterlogged condition

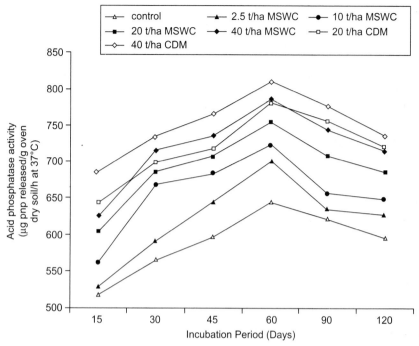

Fig. 9.8   Acid phosphatase activity under different treatments at 60%WHC of soil

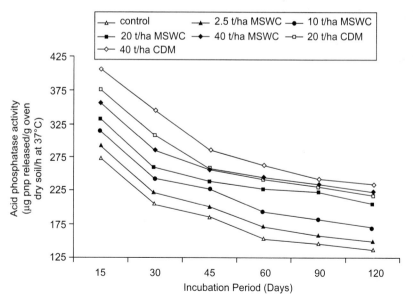

Fig. 9.9   Acid phosphatase activity of soil under different treatments at waterlogged condition

the MSWC-supplemented samples. Enzyme activities, under the WH regime, increased up to 60 days with all treatments and then declined. At 120 days incubation urease activity of soils treated with MSWC recorded either significantly lower values at lower application rates or similar values at higher rates during 15 days of incubation compared to that of the no input treatment. However, CDM treated soils maintained higher urease activity at 120 days compared to the corresponding values at 15 days of incubation. In contrast, at 120 days, phosphatase activity of soils treated with CDM or MSWC at all application rates maintained significantly higher values than at 15 days. Under the WL regime both enzyme activities progressively declined with all treatments, attaining the lowest values at 120 days. Mean periodic variation of urease and acid phosphatase activities gradually increased up to 60 days and then declined. Both enzyme activities were significantly higher in the WH than the WL regime.

The source of enzymes in soil is presumed to originate from microorganisms, plant roots and soil animals (Tabatabai, 1982). Since the soil samples in this study were devoid of extraneous plant and animal materials, it is safe to assume that enzyme activities originated from microorganisms (Gianfreda and Bollag, 1996).

Increased enzyme activities in soil with addition of CDM or MSWC are in agreement with Giusquiani *et al.* (1994) and Perucci (1990). Such increases in enzyme activities were mediated by increases in microbial biomass and hence may be considered a direct contribution of organic inputs (Martens *et al.*, 1992). Elevated levels of enzyme activities due to more organic inputs can also be similarly explained. The periodic variation of enzyme activities of soils at WH regime did not follow the trend of MBC and SR that were found to increase up to 30 days of incubation and subsequently declined. The increased enzyme activities up to 60 days, even after the decrease in MBC, seemed to be related to active enzymes within dead cells, associated cell fragments and activity of living cells (Skujins, 1976). The decline of urease activity and the maintenance of higher phosphatase activity at 120 days incubation corresponding to the values at 15 days project the differential behavior of "protective capacity" of soil due to these two enzymes or it might be the continued production of phosphatase by the existing microorganisms. While the reverse phenomenon occurred in the case of urease. Lower enzyme activity in WL, as compared to the WH regime, agrees with the earlier observation of Pulford and Tabatabai (1988). The decrease in enzyme activities in the soil with or without treatments probably resulted from the increase in reduced metals in WL soils.

Field Study

The pattern of variation of urease and acid phosphatase activities in soil for various treatments and different periods were similar in the three years of study (Figures 9.10 and 9.11). The significantly highest enzyme activities in soil were found in the CDM+Urea plots and the lowest in the control plots. Decomposed cow manure, either alone or with urea, increased the activities of soil enzymes more than applications of MSWC. The application of fertilizer resulted in lower activities of both enzymes than the application of either CDM or MSWC. Urease activity was highest at 15 DAT than the other

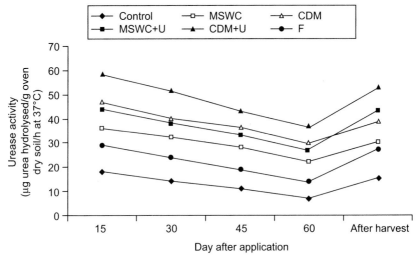

Fig. 9.10 Temporal variation of urease activity in rice soil under different treatments in 1997

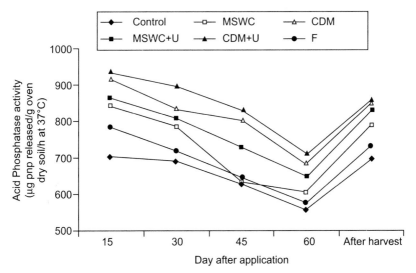

Fig. 9.11 Temporal variation of acid phosphatase activity in rice soil under different treatments in 1997

periods. Generally, both soil enzyme activities started to decline from 15 DAT until 60 DAT. After crop harvest, these activities increased again.

The urease and acid phosphatase activities of soil were measured as these are important soil enzymes in terrestrial N and P cycles. They are also known for their inhibition under conditions of heavy metal pollution (Perucci, 1992). Soil contains both acid and alkaline phosphatases; acid phosphatase predominates in acid soil while alkaline phosphatase in alkaline soil (Eivazi and Tabatabai, 1977). The soil of this experimental site was acidic; hence acid phosphatase activity was measured.

The sources of enzymes in soil are still obscure but are presumed to be related to microbial biomass, plants and animals (Tabatabai, 1994). The contribution of a single factor to the total soil enzyme activity is difficult to resolve. However, evidence could be obtained from this and other investigations (Tyler, 1981) that urease and acid phosphatase activities were positively related to microbial biomass.

The addition of organics (CDM or MSWC) increased enzyme activities of soil. This could have originated from the higher amounts of endoenzymes in the viable microbial populations and increased levels of accumulated enzymes in the soil matrix. The enzymes in the organics-amended soils (Dick and Tabatabai, 1984) may also directly contribute to enzyme activities. Higher enzyme activities of soil treated with CDM rather than MSWC was associated to their capacity for microbial biomass production. Mulvaney and Bremner (1981) earlier reported that the urease activity of soil depended on the type and amount of organic materials applied. Martens et al. (1992) observed that the variation in the nature of the organic materials variably stimulated the production of acid phosphatase and urease in soil.

CONCLUSIONS

The quality parameters of the compost samples showed that the compost was a mature and stable product. Most of the relevant characteristics of compost samples conformed to the standards of quality compost for safe use in agriculture. The major objection in the use of MSWC as a soil supplement has been the possibility of the detrimental influence of heavy metals content on soil quality indicators like MBC, SR and enzyme activities. In the laboratory incubation study, we have demonstrated that the application of MSWC, even at a rate as high as 40 t ha^{-1}, does not negatively impact soil quality parameters. The field study also lends support to evidence that there was no apparent detrimental influence on the soil quality indicators due to short-term application of MSWC in realistic doses, to rice paddies. The present investigation shows that soil quality is enhanced upon the application of MSWC as an ameliorating agent. Therefore, it is proposed that under such conditions where traditional organic supplements are scarce, MSWC can be safely used. However, long-term monitoring of changes in soil quality parameters under field condition is necessary.

REFERENCES

Alexander, M. (1977). Introduction to soil microbiology. 2nd Edn. Wiley Eastern Limited, New Delhi.

Anderson, T.H., Domsch, K.H. (1980). Quantities of plant nutrients in the microbial biomass of selected soils. Soil Science 130: 211-216.

Bhattacharyya, P., Ghosh, A.K., Chakraborty, A., Chakrabarti, K., Tripathy, S., Powell, M.A. (2003a). Arsenic uptake by rice and accumulation in soil amended with municipal solid waste compost. Communication in Soil Science and Plant Analysis 34: 2779-2790.

Bhattacharyya, P., Chakraborty, A., Chakrabarti, K. (2003b). Residual effect of municipal solid waste compost on microbial biomass and activities in mustard growing soil. Archives of Agronomy and Soil Science 49: 585-592.

Bhattacharyya, P., Chakrabarti, K., Chakraborty, A. (2003c). Effect of municipal solid waste compost on microbiological and biochemical soil quality. Compost Science and Utilization 11: 220-227.

Bhattacharyya, P., Pal, R., Chakraborty, A., Chakrabarti, K. (2001). Microbial biomass and its activities of a laterite soil amended with municipal solid waste compost. Journal of Agronomy and Crop Science 187: 207-211.

Bhattacharyya, P., Pal, R., Chakrabarti, K., Chakraborty, A. (2004). Effect of composting on extractability and relative availability of heavy metals present in Calcutta municipal solid waste. Archives of Agronomy and Soil Science 50: 181-187.

Brookes, P.C. (1995). Estimation of adenylate energy charge in soils. In: Methods in applied soil microbiology and biochemistry. p. 204. Eds. K. Alef, P. Nannipieri. Academic Press, London.

Brookes, P.C., Heijen, C.F. McGrath, S.P., Vance, I.D. (1986). Effect of metal toxicity on the size of the soil microbial biomass. Soil Biology and Biochemistry 18: 383-388.

Chakrabarti, K., Bhattacharyya, P., Chakraborty, A. (2003). Assessing of compost quality by stability and maturity parameters. Science and Culture 69: 262-265.

Chanyasak, V., Hirai, M., Kubota, H. (1982). Changes in chemical components and nitrogen transformation in water extracts during composting of garbage. Journal of Fermentation Technology 60: 439-446.

Dick, W.A., Tabatabai, M.A. (1984). Kinetic parameters of phosphatase in soils and organic waste materials. Soil Science 137: 7-15.

Eivazi, F., Tabatabai, M.A. (1977). Phosphatase in soils. Soil Biology and Biochemistry 9: 167-172.

Epstein, E. (1997). The science of composting, pp. 383-415. Technomic Publishing Company, Boston.

Forster, J.C., Zech, W., Wurdinger, E. (1993). Comparison of chemical and microbiological methods for the characterization of the maturity of compost from contrasting sources. Biology and Fertility of Soils. 16: 93-99.

Gallardo-Lara, F., Nogalles, R. (1987). Effect of the application of town refuse compost on the soil plant system - A review. Biological Wastes 19: 35-62.

Ghosh, A.K., Bhattacharyya, P. (2004) Arsenate sorption by reduced and reoxidised rice soils under the influence of some organic matter amendments. Environmental Geology 45: 1010-1016.

Ghosh, A.K., Bhattacharyya, P., Pal, R. (2004). Effect of arsenic contamination on microbial biomass and its activities in arsenic contaminated soils of Gangetic West Bengal, India. Environment International. 30: 491-499.

Gianfreda, L., Bollag, J.M. (1996). Influence of natural and anthropogenic factors on enzyme activity in soil. In: Soil Biochemistry. pp. 123-193. Eds. B. Stotzky, J.M. Bollag. Marcel Dekker Inc. New York.

Giusquiani, P.L., Pagliai, G.P. Businelli, D., Benetti, A. (1994). Urban waste compost: Effect on physical, chemical and biochemical soil properties. Journal of Environmental Quality 24: 175-182.

Harada, Y., Inoko, A. (1980). The measurement of the cation-exchange capacity of composts for the estimation of the degree of maturity. Soil Science and Plant Nutrition 26: 127-34.

He, X.T., Traina, S.J., Logan, T.J. (1992). Chemical properties of municipal solid waste compost. Journal of Environmental Quality 21: 318-29.

Jenkinson, D.S., Ladd, J.N. (1981). Microbial biomass in soil; Measurement and turnover. In: Soil Biochemistry. pp. 415-471. Eds. E.A. Paul, J.N. Ladd. Marcel Dekker Inc., New York.

Jime'nez, I.E., Garcia, P.V. (1989). Evaluation of city refuse compost maturity: A Review. Biological Wastes 27: 115-42.

Joergensen, R.G. (1995). Microbial biomass. In: Methods in applied microbiology and biochemistry. pp. 382-386. Eds. K. Alef, P. Nannipieri. Academic Press, London.

Kanwar, J.S. (1976). Soil Fertility – Theory and Practice. pp. 128-154. Indian Council of Agricultural Research, New Delhi.

Leita, L., De Nobilli, M. Mondini, C. Muhlbacova, G. Marchiol, L. Bragato, G., Contin, M. (1999). Influence of inorganic and organic fertilization on soil microbial biomass, metabolic quotient and heavy metal bioavailability. Biology and Fertility of Soils 4: 371-376.

Martens, D.A., Johanson, J.B., Frankenberger (Jr), W.T. (1992). Production and persistence of soil enzymes with repeated addition of organic residues. Soil Science 53: 53-61.

Mitra, A., Bhattacharyya, P., Chattopadhyay, D.J., Chakraborty, A., Chakrabarti, K. (2003). Physico-chemical properties, heavy metals and their relations in cultivated landfill soils dumped with municipal solid wastes. Archives of Agronomy and Soil Science 49: 163-170.

Mondini, C., Sanchez-Monedero, A., Leita, L., Bragato, G., De Nobili, M. (1997). Carbon and ninhydrin-reactive nitrogen of the microbial biomass in rewetted compost samples. Communication in Soil Science and Plant Analysis 28: 113-22.

Mulvaney, R.L., Bremner, J.M. (1981). Control of urea transformations in soils. In: Soil Biochemistry. pp. 153-196. Eds. E.A. Paul, J.N. Ladd. Marcel Dekker Inc., New York.

Pal, R., Bhattacharyya, P. (2003). Effect of municipal solid waste compost on seed germination of rice, wheat and cucumber. Archives of Agronomy and Soil Science 49: 407-414.

Parr, J.F., Hornick, S.B. (1992). Utilization of municipal wastes. In: Microbial Ecology. pp. 545-559. Eds. F.B. Metting (Jr). Marcel Dekker Inc., New York.

Pascual, J.A., Garcia, C., Hermandez, T., Ayuso, M. (1997). Changes in the microbial activity of an arid soil amended with urban organic wastes. Biology and Fertility of Soils 24: 429-434.

Perucci, P. (1990). Effect of the addition of municipal solid waste compost on microbial biomass and enzyme activities in soil. Biology and Fertility of Soils 10: 221-226.

Perucci, P. (1992). Enzyme activity and microbial biomass in a field soil amended with municipal refuse. Biology and Fertility of Soils 14: 54-60.

Ponnamperuma, F.N. (1972). The chemistry of submerged soils. Advances in Agronomy 24: 26-29.

Pulford, I.D., Tabatabai, M.A. (1988). Effect of waterlogging on enzyme activities in soil. Soil Biology Biochemistry 20(2): 215-218.

Roy, S., Bhattacharyya, P., Ghosh, A.K. (2004). Influence of toxic heavy metals on activity of acid and alkaline phosphatase enzymes in metal contaminated landfill soils. Australian Journal of Soil Research 42: 339-344.

Sakamoto, K., Oba, Y. (1991). Relationship between the amount of organic material applied and soil biomass content. Soil Science and Plant Nutrition 37: 387-397.

Saviozzi, A., Levi-Minzi, R., Riffaldi, R. (1988). Maturity evaluation of organic waste. Biocycle 29: 54-56.

Skujins, J.J. (1976). Extracellular enzymes in soil CRC. Critical Reviews in Microbiology. p. 382-421.

Sparling, G.P. (1985). The Soil biomass. In: Soil Organic matter and biological activity. p. 223. Eds.D.Vaughan, R.E. Malcolm. Martinus Nijoff/ Dr. W. Junk, Dordrecht.

Stratton, M.L., Barker, A.V., Rechcigl, J.E. (1995). Compost. In: Soil amendments and environmental quality. pp 249-09. Eds. J.E. Rechcigl. Lewis Publishers, London.

Tabatabai, M.A. (1982). Soil enzymes. In: Methods of Analysis. p. 903-947. Eds. A.L.Page *et al.* Part 2. 2nd Wd. Series Agronomy 9. ASA-SSSA, Madison.

Tabatabai, M.A. (1994). Soil enzymes. In: Methods of soil analysis. pp. 775-826. Eds. R.W. Weaver, *et al.* Part 2. Soil Science Society of America, Madison.

Tate, R.L. (2000). Soil Microbiology. 2nd Edn. John Wiley & Sons Inc. New York.

Tyler, G. (1981). Heavy metals in soil biology and biochemistry. In: Soil Biochemistry. pp. 371-413. Eds. E.A. Paul, J.N. Ladd. Marcel Dekker Inc., New York.

Vance, E.D., Brookes, P.C., Jenkinson, D.S. (1987). An extraction method for measuring soil microbial biomass C. Soil Biology and Biochemistry 19: 703-707.

Witt, C., Biker, U., Galicia, C.C., Ollow, J.C.G. (2000). Dynamics of soil microbial biomass and nitrogen availability in a flooded rice soil amended with different C and N sources. Biology and Fertility of Soils 30: 520-527.

Zucconi, F., De Bartoldi, M. (1987). Compost specifications for the production and characterization of compost from municipal solid waste. In: Compost production, quality and use. pp 30-50. Eds. M.De Bertoldi, *et al.* Elsevier, London.

CHAPTER 10

Phytoextraction of Lead-Contaminated Soils: Current Experience

John Pichtel

Ball State University, Natural Resources and Environmental Management, Muncie, IN 47306, USA.

Lead (Pb) is a common contaminant in soils that have been amended with biosolids, exposed to industrial atmospheric emissions, and received industrial and other wastes. A variety of innovative technologies are available for the remediation of lead-contaminated soils; however, many are capital- and technology-intensive, and some pose detrimental effects to the chemical and physical properties of the soil under treatment. As an alternative, the less expensive, more environmentally benign use of green plants as a remediation tool for metal-contaminated soils ("phytoextraction") is being examined. Although phytoextraction is a relatively new application of site restoration, several green plants have been identified with the capability of removing substantial amounts of toxic metals from soil. Phytoextraction technologies are not a panacea for all metal-affected soils, however; metals such as lead may be difficult for plants to remove and translocate. The current goals of phytoextraction are to formulate innovative, economical and environmentally compatible techniques to remove metals from soils, sediments and other environmental media. This chapter will examine the current state of the science of lead phytoremediation technologies. Included are assessment of promising plant types, Pb dynamics in soils, and methods to mobilize soil Pb. Case studies are also provided.

INTRODUCTION

Numerous sites in industrialized nations are contaminated with lead and other toxic metals. Lead may be released into soils from agriculture (for example, from application of lead-enriched biosolids) and mining (via tailings and fugitive emissions). Other lead sources include lead smelting and refining, gasoline additives, paint removal, pesticide production, automobile demolition, and lead-acid battery disposal. All these sources may result in significant detrimental effects to public health and local environments.

The global background level of lead in surface soils is reported to be 20 mg/kg (Kabata-Pendias, 2001). The ionic species of lead in the biosphere are typically Pb^{2+} and, to a lesser degree, Pb^{4+}. Lead and other transition metals exist in soils as dissolved ions in the soil solution or bound to a specific phase such as the lattice of crystalline minerals, interlayer positions of clays, adsorption sites on mineral surfaces, oxides and hydroxides, and complexation sites on organic matter. A schematic showing the various lead reactions in soil appears in Fig. 10.1.

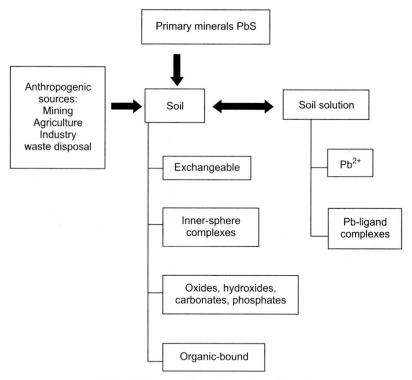

Fig. 10.1 Lead transformations in soil

Lead adsorption onto charged soil particle surfaces results from the formation of relatively weak outer sphere complexes through cation exchange reactions, or the formation of strongly bound inner sphere complexes through ligand exchange reactions (Evans, 1989). Adsorption via cation exchange reactions is not an important mechanism in the retention of metals in heavily contaminated soils. This is supported by studies that have reported extremely low exchangeable metal fractions in metal-contaminated soils (Elliot and Brown, 1989; Ramos et al., 1994; Van Benschoten et al., 1994; Yarlagadda et al., 1995; Chlopecka et al., 1996; Heil et al., 1996; Ma and Rao, 1997; Pichtel et al., 2000). Lead may also complex with soil organic matter (humic substances) and be strongly retained. Humic substances complex with metals via their relatively high content of carboxyl and phenolic groups, and to a lesser extent imino, amino, and thiol groups (Kerndorff and Schnitzer, 1980; Stumm and Morgan, 1996).

In highly contaminated soils, the metal sorption capacity is often exceeded and precipitation of metals as secondary minerals may occur. The most important of the lead precipitates are the oxides, hydroxides, oxyhydroxides, and carbonates. Also, sulfide minerals may form under reducing conditions via interaction of lead with sulfide compounds produced by microbially-mediated SO_4^{2-} reduction (Evans, 1989). Yarlagadda et al. (1995), using scanning electron microscope and energy dispersive X-ray analyses, observed lead-bearing particles from a waste site soil occurring as either metallic lead or a lead precipitate. Lead carbonate ($PbCO_3$), hydrocerussite ($Pb_3(CO_3)_2(OH)_2$), and leadhillite ($Pb_4SO_4(CO_3)_2(OH)_2$) are likely to form in carbonate soils (Royer et al., 1992). Reed et al. (1995) determined that anglesite ($PbSO_4$) was a common lead species at battery disposal sites. Hessling et al. (1990) observed that $PbSO_3$, $PbSO_4$ and lead oxide (PbO_2) were the dominant lead forms in contaminated soil from battery recycling sites. Pichtel et al. (2001), using x-ray diffraction techniques, revealed the presence of both $PbSO_4$ and metallic lead in a Superfund soil (Fig. 10.2).

As a result of the above reactions, lead is considered to be the least mobile of the heavy metals. After being released into soil, only very small proportions will migrate into groundwater or surface water. Contamination of soils with lead is mainly an irreversible and, therefore, a cumulative process in surface soils (Smith et al., 1995).

Conventional remediation methods for lead-contaminated soil include soil washing, solidification/stabilization, and excavation. Many of these technologies are costly and involve the application of sophisticated equipment and training. Some only serve to isolate the contaminant rather than remove it. Additionally, some technologies pose detrimental effects to the chemical and physical properties of the soil under treatment. As an alternative, the less expensive, more environmentally benign use of green plants as a remediation tool ("phytoremediation") for lead-contaminated soils is being examined at the greenhouse scale and in the field.

The two principal strategies in use in metal-based phytoremediation are phytostabilization and phytoextraction. The goal of phytostabilization is to rapidly reduce metal mobility in soil, utilizing selected plants to transform

Fig. 10.2 Scanning electron micrograph of anglesite (PbSO$_4$) and metallic lead in a Superfund soil

metals to less toxic forms by precipitation or other processes. Plant cover will also reduce leaching, runoff, and erosion via stabilization of soil by roots. Stabilization can be augmented by adding various reagents to the soil (for example limestone or phosphate fertilizer) to reduce metal mobility. It must be emphasized that phytostabilization will maintain the metal concentration in the soil. Phytostabilization will not be addressed in any length in this chapter.

The opposite end of the phytoremediation spectrum involves the accumulation of metals in shoot tissue for eventual removal and treatment (phytoextraction). The ability to accumulate metals may have originally evolved as a defense against plant-eating insects or bacterial or fungal diseases (Baker and Brooks, 1989). The process initially involves uptake of a soluble metal by the plant root. In the case of lead, the soluble fraction (Pb^{2+}) is absorbed primarily by root hairs (Kabata-Pendias, 2001). Such uptake can be passive or occur via attachment to specific carrier molecules produced by the root. Once carried across the root membrane, the metal may remain in the root or it may be translocated to upper plant parts (Fig. 10.3). Transpiration may be the major driving force that directs lead accumulation in plant shoots (Blaylock *et al.*, 1997). The metal-enriched plant tissue is harvested, either repeatedly over the growing season or during a single event at the end of the season. The harvested plant material may be used for non-food purposes. It can also be processed via chemical or thermal techniques to recover the metal or it can be shipped to an appropriate facility for disposal.

In order for phytoextraction to be both technically feasible and an economically viable technology, target plants must tolerate, absorb and translocate concentrations of lead and other environmentally important metals in concentrations of 1% or greater (Huang *et al.*, 1997). Plants also should produce sufficient biomass to allow for efficient harvesting and disposal of metal-rich tissue. Finally, plants must be responsive to agricultural practices that allow for repeated planting and harvesting in contaminated soil (Baker *et al.*, 1994; Kumar *et al.*, 1995; Blaylock *et al.*, 1997).

Phytoextraction can be used in conjunction with other cleanup methods (e.g., as a final, polishing step at a partly cleaned site). Much research is on-going in this innovative application of soil cleanup. To enhance the efficiency of metal phytoextraction, a range of practical strategies are being investigated, including: (a) identification of suitable plant species via screening studies, (b) optimization of agronomic practices for maximizing biomass production and metal uptake (Chaney *et al.*, 2000), and (c) modification of species via conventional breeding or genetic engineering (Bennett *et al.*, 2003). There have been limited examples of phytoremediation application in field tests; the technique is still in the developmental stage (Banuelos, 2000; Blaylock, 2000; Shen, 2002).

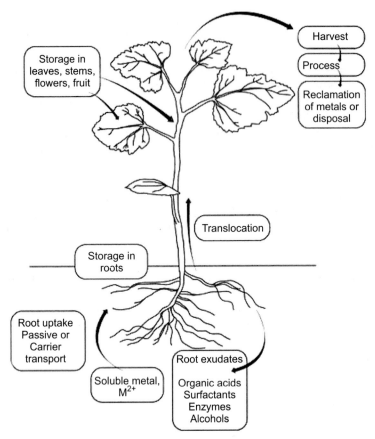

Fig. 10.3   Lead uptake and translocation in green plants

PLANT TYPES FOR LEAD EXTRACTION

The selection of appropriate plant species is one of the most critical factors affecting the degree of metal removal at a site. In order to optimize cleanup of contaminated soils via phytoextraction, it is essential to identify plants that tolerate and absorb high levels of metals, preferably into above-ground biomass. The rate of metal removal depends upon the biomass harvested and metal concentration in harvested biomass. One subject of practical concern relates to whether the remediating species should be a metal hyperaccumulator or a common non-accumulator species. Hyperaccumulator plants have the potential to concentrate high metal levels in their tissue; however, their use may be limited by small size and slow growth. In non-accumulator species, the low potential for metal concentration may be compensated by the production of significant biomass (Ebbs *et al.*, 1997; Lasat, 2000).

Various plant types including trees, vegetable crops, grasses and weeds are known to accumulate a wide range of heavy metals. There are estimated to be several hundred known so-called metal hyperaccumulators world-wide; however, only a limited number of these hyperaccumulate lead. The potential for metal extraction is of primary importance; however, other criteria such as ecosystem stability must be also considered when selecting plants for phytoextraction. Native species tend to be preferred to exotic plants as the latter can be invasive and endanger the integrity of the ecosystem. To avoid propagation of weedy species, crop plants are typically preferred, although some may be too palatable and pose a risk to grazing animals (Lasat, 2000). Plants identified with the natural potential to uptake lead often belong to the families Brassicaceae, Euphorbiaceae, Asteraceae, Lamiaceae, and Scrophulariaceae. The background concentration of lead in plant tissue is 10 mg g^{-1} and lead hyperaccumulating plants are defined as those in which lead concentrations in the dry matter of leaves exceed 1000 mg kg^{-1} (Baker and Brooks, 1989). This class of plants is often found at metalliferous sites (Shen *et al.*, 2002). Table 10.1 lists some lead accumulating plants identified in recent years.

Table 10.1   Selected lead accumulating plants

Scientific Name	Common Name
Apocynum cannabinum	Hemp dogbane
Armeria maritima	Seapink thrift
Ambrosia artemisiifolia	Ragweed
Brassica juncea	Indian mustard
Brassica napus	Rape
Brassica oleracea	Cabbage
Carduus nutans	Nodding thistle
Commelina communis	Asiatic dayflower
Festuca ovina	Blue/sheep fescue
Helianthus annuus	Sunflower
Thlaspi rotundifolium	Pennycress
Triticum aestivum	Wheat
Zea mays	Maize

Sources: Reeves and Brooks, 1983; Pichtel *et al.*, 2000; Berti and Cunningham, 2000; U.S. EPA, 2000a

Measurement of lead removal efficiency by a green plant can be assessed in several ways; first, the total lead content of plant tissue may be measured after reaction with a suitable digesting solution (e.g., aqua regia) followed by atomic absorption spectrophotometry. Such data can be deceiving, how-ever. A lead hyperaccumulator that does not produce significant biomass is of little use in a field situation. To address this weakness, tissue lead concentrations can be multiplied by the total plant dry matter production, thus calculating total lead removal per unit mass or volume of soil. Alternatively, a phytoextraction coefficient can be calculated. This coefficient is the ratio of the metal concentration found within above-ground biomass to the metal concentration found in the soil. Thus, the greater the coefficient, the greater the uptake of contaminant (U.S. EPA, 2000a).

The most frequently cited lead hyperaccumulators are cultivars of Indian mustard (*Brassica juncea*). This species not only absorbs lead through the roots, but it also has the ability to translocate lead from roots to shoots, which is a critical characteristic for efficient phytoextraction. When grown in a lead-enriched nutrient solution, up to 1.5 % Pb^{2+} was found in shoot tissues (Kumar *et al.*, 1995). The phytoextraction coefficient for Indian Mustard is 1.7 (U.S. EPA, 2000b). Translocation to shoots in Indian mustard, however, may be restricted when grown in soils where Pb^{2+} bioavailability is limited. Therefore, in certain situations it may be necessary to harvest the roots as well as the shoots of these plants. Increasing the bioavailability of soil lead is covered later in this chapter. *Thalspi rotundifolium* ssp. Cepaeifolium (Pennycress) is a non-crop Brassica which has been found to grow in mine soils enriched with lead and zinc. *Thalspi rotundifolium* (L.) Gaud.-Beaup can achieve a shoot concentration of 8500 mg g^{-1} (Reeves and Brooks, 1983). Unfortunately, this Thalspi species is a low biomass producer and has a slow growth rate, which makes it unsuitable for phytoextraction.

Because many hyperaccumulator species are not suitable for phytoremediation in the field due to low biomass production and slow growth, it has been suggested to use high biomass species such as maize (*Zea mays* L.), pea (*Pisum sativum* L.), oat (*Avena sativa* L.), canola (*Brassica napus* L.), and barley (*Hordeum vulgare* L.), with improved plant and soil management practices to enhance metal uptake by these species (Huang and Cunningham, 1996; Banuelos *et al.*, 1997; Blaylock *et al.*, 1997; Huang *et al.*, 1997; Ebbs and Kochian, 1998; Ajwa *et al.*, 1999; Shen *et al.*, 2002). Bench-scale studies have shown that certain crop plants are capable of promising rates of phytoextraction. Maize, alfalfa, sorghum, and sunflower (*Helianthus annuus*) were found to be effective due to their rapid growth rate and substantial quantities of biomass produced (U.S. EPA, 2001a). The highest bioaccumulation of lead among crop plants is reported for leafy vegetables (especially lettuce) grown near nonferrous metal smelters where plants are exposed to lead sources in both soil and air. In these locations, lettuce has been found to contain up to 0.15% Pb (dry weight) (Smith *et al.*, 1995). Lead has also been found to accumulate in certain crops such as oats grown on biosolids-amended soils (Pichtel and Anderson, 1997). Other high biomass producers with the ability to uptake lead include common weeds such as ragweed (*Ambrosia artemisiifolia*) (Pichtel *et al.*, 2000).

AGRONOMIC PRACTICES FOR ENHANCING PHYTOEXTRACTION

As with production of any crop plant, metal-accumulating plants will respond favorably to establishment of optimum soil chemical and physical characteristics and plant husbandry. For example, the application of the appropriate fertilizer elements in the proper proportions will increase biomass production. However, some precautions are noteworthy. For example, the addition of phosphorus fertilizer during phytoextraction can inhibit the uptake of lead due to precipitation as pyromorphite and chloro-pyromorphite (Chaney *et al.*, 2000). Foliar application of P is one means to circumvent this problem.

Planting practices will also affect success of metal removal. The extent of metal extraction depends in large part on the total quantity of biomass produced. Plant density (number of plants m^{-2}) affects biomass production as it affects both yield per plant and yield ha^{-1}. In general, a higher planting density tends to minimize yield per plant and maximize yield per hectare. Density may also affect the pattern of plant growth and development. For example, at higher stand density, plants will compete more aggressively for light. As a result, more nutrients and energy may be allocated for plant growth as opposed to developmental processes (e.g., flowering and reproduction). Additionally, the distance between plants may affect the structure of the root system with subsequent effects on metal uptake (Lasat, 2000).

Proliferation of weeds and diseases may significantly decrease yields; therefore crops used for soil remediation must be rotated. If phytoremediation is anticipated to last for a short period (e.g., two to three years), monoculture may be an acceptable practice. However, for longer-term applications, as is the case for most metal phytoextraction projects, successful metal cleanup probably will not be achieved with only one species used in monoculture. Plant rotation is even more important when multiple crops per year are to be grown (Lasat, 2000).

SOLUBILIZATION OF SOIL METALS FOR PHYTOEXTRACTION

When lead is applied in soluble form in hydroponic nutrient solutions, roots of several plant species are capable of taking up substantial amounts; the rate of uptake increases with higher solution concentrations and with time. The situation becomes more complex and difficult, however, when lead uptake from *soil* is considered. Only the most effective lead hyperaccumulators accumulate a Pb^{2+} concentration greater than 0.1% in their shoots when grown in contaminated soils without the addition of specialized amendments (Huang *et al.*, 1997). The transfer of lead and other heavy metals from soil to plant is dependent on three factors: (a) the total amount of potentially available metal; (b) the activity of elements in the soil solution; and (c) the rate of element transfer from solid to liquid phases and to plant roots (Brümmer *et al.*, 1996; Schmidt, 2003). Because effective phytoextraction is dependent on the above factors to achieve significant uptake, soil conditions may need to be altered to increase metal solubility and availability (Henry, 2000).

Synthetic Chelating Agents to Enhance Phytoextraction

The two primary amendment techniques used to increase lead bioavailability in soils and mobility within the plant are adding synthetic chelates and lowering soil pH. Chelating agents form water-soluble chelate-Pb complexes, thus preventing lead precipitation and rendering lead available for uptake into roots and transport within the plant. Synthetic chelating agents such as ethylethylenediaminetetraacetic (EDTA), hydroxyethylethylenediamine-triacetic acid (HEDTA), diethylenetriaminepentaacetic acid (DTPA) and nitrilotriacetic acid (NTA) (Fig. 10.4) have been used in pot and field experiments to enhance heavy metal uptake by plants (Huang and Cunningham, 1996; Blaylock *et al.*, 1997; Huang *et al.*, 1997; Ebbs and Kochian, 1998; Kayser *et al.*, 2000; Bricker *et al.*, 2001). Others include cyclohexylene-dinitriloetetraacetic acid, ethyleneglycol-bis(β-aminoethyl ether) *N,N,N'*

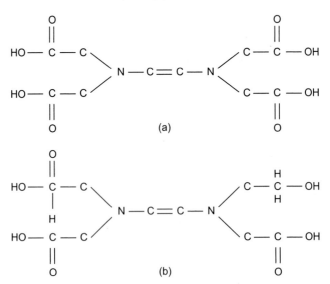

Fig. 10.4 Synthetic chelating agents (a) ethylethylenediaminetetraacetic acid (EDTA), (b) hydroxyethylethylenediaminetriacetic acid (HEDTA)

Fig. 10.4   (c) diethylenetriaminepentaacetic acid (DTPA) and (d) nitrilotriacetic acid (NTA)

N-tetraacetic acid, EGTA (ethylenebis [oxyethylenetrinitrilo] tetraacetic acid) and ethyleneidamine-dissuccinate (Grèman *et al.*, 2003).

If hyperaccumulation of soil lead is to be augmented, the choice of appropriate chelating agents is one of the first issues to address. Predictions regarding the ability of a chelating agent to solubilize a metal in the soil can be made using equilibrium constants found in the literature and assuming that metal solubility is controlled by known solid phases. However, the variability associated with contaminant form as well as soil properties at contaminated sites makes it difficult to accurately predict all reactions and conditions that influence lead-to-chelant binding. Empirical data is therefore essential to determine the effectiveness of an applied chelating agent to increase metal solubility (Henry, 2000).

A range of plant types have been found to effectively uptake soil lead after application of chelating agents. In a study by Chaney *et al.* (1997), the highest lead transfers from the soil solution to shoots were found in Indian mustard and pea, followed by maize and sunflower. Maize has shown more than a twenty-fold increase in the amount of lead concentrated in shoot tissue when grown in soil amended with EDTA and HEDTA. The amount of lead translocated to shoots was proportional to the amount of chelating agent added to soil (Chaney *et al.*, 2000). Lead concentrations of 4500 mg kg^{-1} (45-fold the control concentration) were achieved when maize plants were transplanted to lead-enriched soil 10 days after germination. In studies by Huang *et al.* (1997) and Epstein *et al.* (1999), soil-bound lead was solubilized by EDTA and DTPA in a short period (6 hours), and the majority of lead uptake by the plants may have occurred rapidly after EDTA application.

Multiple chelant applications are known to affect lead uptake differently as compared with a single application. Grèman *et al.* (2001) studied the effect of agent application frequency on lead accumulation efficiency. They found that a single dose of 2.9 g EDTA kg^{-1} enhanced lead accumulation of cabbage (*Brassica oleracea* L.) grown in a greenhouse 105-fold, as compared with a 44-fold increase if the same amount of EDTA was split and added in four dosages. In contrast, in the study of Puschenreiter *et al.* (2001), the lead concentration in maize grown on a soil with 5600 mg total Pb kg^{-1} increased eightfold to 49 mg kg^{-1} after adding 1 g EDTA kg^{-1} three weeks before harvesting; this value increased to 18-fold when the EDTA application was split into three dosages over a period of three weeks.

Several studies have shown that once chelates are assimilated by the root, they are transferred nearly entirely from roots to shoots (Barber and Lee, 1974; Hamon *et al.*, 1995; Vassil *et al.*, 1998). Labeling the carbon fraction of EDTA revealed that plants can take up entire metal-EDTA complexes, which are subsequently transported in the xylem sap (Fig. 10.5) (Schmidt, 2003; Vassil *et al.*, 1998).

Additional Considerations in Chelate Use

Although beneficial to lead solubilization and uptake, chelate induced-hyperaccumulation may actually prove fatal to the plant. Of course, the dead plant tissue can still be harvested and processed for lead recovery or disposal.

An additional consideration with chelant applications pertains to the entry of lead to the food chain, thus increasing bioavailability to animals and humans. Increasing lead solubility using chelating agents increases the chance for the metal to leach out of the soil profile and enter groundwater. Additionally, chelates and chelate-lead complexes may be toxic to indigenous soil microorganisms. Therefore, the toxicity of the chelating agents and their metal complexes in soils need careful assessment (Means *et al.*, 1980; Borgmann and Norwood, 1995; Nortemann, 1999; Grèman *et al.*, 2001). To avoid possible metal-chelate movement into groundwater and detrimental effects of the remaining chelant on soil microorganisms, the amount applied and method of chelate application are important. Innovative irrigation techniques and time control of chelate application can also be utilized (Shen *et al.*, 2002).

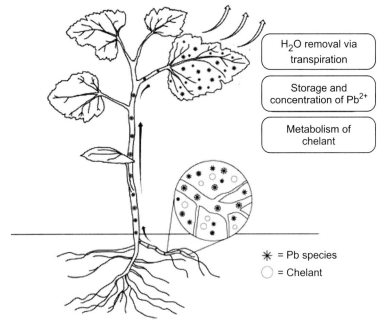

Fig. 10.5   The effect of chelate addition to soil on lead uptake in a green plant

Certain organic wastes may prove suitable for heavy metal chelation and solubilization in soils. Pichtel and Anderson (1997) found that the addition of composted sewage sludge (CSS) to soil converted some soil metals to fractions which were more plant-available (e.g., soluble and exchangeable forms of Pb^{2+}). The addition of CSS as a chelating agent and soil conditioner could possibly be employed as part of a phytoextraction strategy (Bricker *et al.*, 2001).

Soil pH Effects

pH is one of the principal soil factors controlling metal availability and is thus significant for metal uptake by plants. By maintaining a moderately acidic soil pH, lead bioavailability and plant uptake has been shown to increase (Salt *et al.*, 1995; Huang *et al.*, 1997). Soils can be acidified through the application of ammonium- or sulfur-containing fertilizers, or acidifiers such as elemental sulfur, aluminum sulfate or certain industrial wastes. In a study performed by Chlopecka *et al.* (1996) on metal contaminated soils in southwest Poland, soil samples with a pH less than 5.6 contained relatively more of all soil metals in the exchangeable form than in samples where pH was greater than 5.6. In addition, at lower pH values soil lead had a greater potential to translocate from roots to shoots. The ideal pH range for the growth of most plants is from 5.0-8.0; a pH range of approximately 5.5 to 6.0 may be optimal for phytoextraction, since lower values may inhibit plant growth.

CASE STUDIES OF PHYTOEXTRACTION OF LEAD

In recent years plants having the potential for lead phytoextraction have been tested in the field in limited studies. Much of the work involves some of the more common hyperaccumulators (e.g., Indian mustard), although others have tested high biomass producers such as maize. Synthetic chelating agents have proven essential to enhancing hyperaccumulation. Selected field projects from the United States are reported below.

Bayonne, New Jersey, USA

At an industrial site in Bayonne, New Jersey, soil was contaminated with high lead levels resulting from cable manufacturing operations (Blaylock *et al.*, 1999). Lead concentrations in the surface 0-15 cm of soil ranged from 1,000 to 6,500 mg kg^{-1} with an average of 2,055 mg kg^{-1}. At depths of 15-30 cm, lead concentrations ranged from 780 to 2,100 mg kg^{-1}, and at 30-45 cm depth soil lead measured 280 to 8,800 mg kg^{-1}. The soil was a sandy loam containing 2.5% organic matter and had a pH of 7.9.

A field project was conducted to investigate the use of phytoremediation as a lead removal technology. To enhance the bioavailability of soil lead, EDTA was applied at a rate of 2 mmol kg^{-1} via an irrigation system. Indian mustard was the test crop; three crops were grown and each was harvested after six weeks of growth.

By the end of the growing season, the three crops had reduced the soil lead concentrations from 2,300 to 420 mg kg^{-1} in the surface soil (average 960 mg kg^{-1}). At the 15-30 cm depth average lead concentrations decreased to 992 mg kg^{-1}, and there were very minor decreases in lead concentrations at 30-45 cm depth.

Although none of the soil lead levels at this site were brought below regulatory limits within the first year, the results demonstrate the potential of phytoremediation to reduce lead levels in contaminated soils. An encouraging result was that no leaching of lead or EDTA was observed, thus indicating that the reduction in soil lead concentrations was due to removal by plants and not by leaching (Blaylock *et al.*, 1999).

Magic Marker Site, Trenton, New Jersey, USA

The Magic Marker site was contaminated from the manufacturing of lead acid batteries and other industrial activities (Blaylock *et al.*, 1999). The concentration of soil lead measured as high as 1,800 mg kg^{-1} in places. The majority (72%) of soil lead was assessed as available for plant uptake. Soil pH ranged from 5.1 to 7.1.

In order to reduce the total soil lead levels, phytoextraction was implemented on a 4,500 ft^2 (420 m^2) area of land. Indian mustard was used as the test species. Ethylenediaminetetraacetic acid was applied at a rate of 200 mmol m^{-2}. Three crops were grown in a single season and each was harvested after six weeks of growth. After the third crop was harvested, none of the treated area exceeded 800 mg kg^{-1} and only small areas exceeded 600 mg kg^{-1} by the end of the growing season.

The most significant impact of phytoextraction was on the surface soil (0-15 cm) where the average lead concentration was reduced by 13%. In areas exceeding 600 mg kg^{-1}, phytoextraction decreased lead concentrations from 736 to 539 mg kg^{-1}, a 21% decrease.

In an attempt to determine possible downward movement of lead into the soil profile, one area was treated with 2 mol EDTA m^{-2} in the fall. By the following spring, soil samples were collected to a depth of 120 cm. Total soil lead concentrations decreased with depth; 878 mg kg^{-1} was found at the surface compared to 15 mg kg^{-1} at 120 cm depth. The EDTA remained primarily in the upper 30 cm of the soil (Blaylock et al., 1999).

Open Burn/Open Detonation Area, Ensign-Bickford Company, Simsbury, Connecticut, USA

The OB/OD area at the Ensign-Bickford Company was highly contaminated with lead due to previous weapons-related activities (FRTR, 2000). The soil was a silt loam and pH ranged from 6.5 to 7.5. From 1996-1997, a full-scale phytoremediation project was conducted on 1.5 acres (0.6 ha) of contaminated soil. The area of study was subsequently increased, and both phytoextraction and phytostabilization technologies were utilized. A total of 2.35 acres (0.95 ha) was divided into five treatment areas. The total soil lead concentrations were as shown below:

Area	Pb concentration mg kg^{-1}
1	500–5,000
2	125–1250
3	500–2000
4	750–1000
5	6.5–7.5

Areas 1 through 4 were treated using phytoextraction, and phytostabilization was implemented at area 5. Soil amendments were applied to areas 1 through 4 to increase the mobility of lead within the soil profile. Three crops were planted and harvested during the 1998 growing season. The first crop grown was Indian mustard, the second was sunflower, with the third crop a combination of both species. Soils were fertilized with nitrogen, phosphorus, and potassium. Dolomitic limestone was added to adjust soil pH. Fertilizers and limestone were tilled into the soil to a depth of 15 to 20 cm. An overhead irrigation system was used and supplemental foliar fertilizers were added through the irrigation system.

Plant growth at the site was generally good; however, certain areas were excessively wet, resulting in poor plant growth and decreased biomass production. Phytoextraction in areas 1 through 4 decreased total soil lead concentrations from an initial average of 635 mg kg^{-1} (April 1998) to 478 mg kg^{-1} (October 1998). Lead uptake in Indian mustard ranged from 342 mg kg^{-1} (dry weight) for the first crop to 3,252 mg kg^{-1} for the third crop. The average lead uptake was similar in both sunflower and Indian mustard with a value of 1000 mg kg^{-1} in the sunflower and 1,091 mg kg^{-1} (dry weight) in Indian mustard (FRTR, 2000).

Superfund Site, Indiana, USA

A phytoremediation field study was conducted on a Superfund site measuring approximately 7.5 acres (3 ha), contaminated over 90 years from battery recycling, foundry, and secondary smelting operations resulting in significant soil contamination by lead (Pichtel *et al.*, 2000). Total soil lead concentrations averaged 55,480 mg kg^{-1}, with maximum values reaching 140,500 mg kg^{-1}. It was estimated that 71.4% of the total soil lead was in the non-residual form, thus bioavailable for plant uptake. Soil pH ranged from 7.5 to 8.1. Homes occur immediately adjacent to the site.

Eighty-five percent of the area was vegetated by native plant species, predominantly 'grasses, legumes and assorted perennials. Due to the extremely high levels of total soil lead, roots of all plants were severely stunted, generally not penetrating past a depth of 5 cm. Some common plants identified included *Agrostemma githago*, plantain *(Plantago rugelii)*, garlic mustard *(Alliaria officinalis)*, dandelion *(Taraxacum officinale)*, ragweed *(Ambrosia artemisiifola)*, and red maple *(Acer rubrum)*.

Lead uptake by the plants varied from non-detectable to 1,800 mg kg^{-1} taken up by *Agrostemma githago* roots and 1467 mg kg^{-1} in Black medic *(Medicago lupulina)* shoots. Plants that removed lead were predominantly herbaceous species, with some producing substantial biomass. Of the plants studied, more than 65% had a higher concentration of lead in roots than in shoots.

Dandelion and ragweed were subsequently studied in growth chambers to assess their ability to extract lead from the Superfund soil. Dandelion was successful at extracting 1059 mg kg^{-1} of lead from the contaminated soil in the first crop and 921 mg kg^{-1} in the second. The first crop of ragweed was successful at extracting 965 mg kg^{-1} of lead, with an increase to 1232 mg kg^{-1} for the second crop.

One of the more salient findings of the field assessment at the Superfund site was the ability of the plants to survive in such highly contaminated soils.

Twin Cities Army Ammunition Plant (TCAAP), Minneapolis-St. Paul Minnesota, USA

The Twin Cities Army Ammunition Plant (TCAAP) is a 2,370 acre (960 ha) facility that has been involved in the production of small arms ammunition, fuses and artillery shell materials (FRTR, 2002). In 1981, it was determined that contaminated groundwater from TCAAP was migrating into the Minneapolis-St. Paul metropolitan groundwater supply.

Two sites were investigated, designated site C and site 129-3. Total soil lead concentrations ranged from an average of 2,610 ppm in the surface soil at site C to an average of 358 ppm in the surface soil of site 129-3. The soil at site C was peat underlain by fine sand and sandy clay. Site 129-3 was composed of fine- to medium-grained sand.

Two crops were grown on the sites: Maize was the first crop and white mustard (*Brassica alba*) was the second. Ethylenediaminetetraacetic acid and acetic acid were applied to the soil to enhance lead mobilization and uptake by plants.

The first crop, maize, yielded only 2.1 to 3.6 tons of biomass per acre (4700-8000 kg per ha) compared to the anticipated yield of 6 tons per acre (1340 kg ha^{-1}), and the dry weight lead concentrations averaged 0.65% and 0.13% for sites C and 129-3, respectively. White mustard, the second crop, yielded 1.9 to 2.1 tons of biomass per acre (4260-4700 kg ha^{-1}). Average dry weight lead concentrations were relatively low: 0.083% and 0.034% for sites C and 129-3, respectively (FRTR, 2002).

Overall, the results were not as promising as had been anticipated. However, due to the average annual temperature of approximately 50°F (10°C) at this site (less than ideal for growing many agricultural crops), this project provided an opportunity to examine the feasibility of phytoextraction in non-optimal climatic conditions.

ADVANTAGES AND LIMITATIONS OF PHYTOREMEDIATION

As with most technologies, both advantages and disadvantages are inherent to the use of phytoextraction of contaminated soils. Before scientists, planners, and regulatory officials consider whether or not it would be advantageous to employ a phytoremediation technology, a range of technical, environmental, economic, and social considerations must be assessed.

Advantages

One of the most attractive features of phytoextraction is its cost-effectiveness. Principal costs are generally limited to tilling of the seedbed and the planting, harvesting and disposal of the plant tissue. Routine monitoring and laboratory analysis costs must also be accounted for.

Phytoextraction can be carried out either *in situ* or *ex situ*. *In situ* applications can take place without severely disturbing the site. Because phytoextraction often treats the more mobile and available fraction of a metallic contaminant, it is compatible with risk-based remediation programs that leave the unleachable fraction in the soil.

Remediation of a contaminated site with plants tends to be more aesthetically pleasing and therefore more acceptable to the public (Glass, 2000). Additionally, green plants offer a degree of versatility that may not be possible with conventional remediation technologies. For example, plants may be well suited for a very saturated soil that restricts the use of heavy machinery (Pichtel, 2000).

Disadvantages

The most significant disadvantage of phytoextraction technologies is time. Growing season and plant life cycles are difficult to significantly alter. Additionally, some hyperaccumulators are notoriously slow growing. For this reason, phytoremediation is not suitable for contaminated sites that pose an immediate health risk. Additionally, the heterogeneous nature of some soils may require that several plant species be grown at a site, possibly in succession, further increasing the amount of time needed for the remediation to be successful (Pichtel, 2000).

The use of plants, like all remediation methods, does not allow for 100% removal of contamination. As concentrations of contaminants change, the physico-chemical interactions between soil, plant, contaminants, and the associated environment change. Additionally, most hyperaccumulators have relatively shallow root masses and are therefore only practical for remediation of the topmost meter of soil (Glass, 2000).

The use of invasive, nonnative species has the potential to affect biodiversity. Unfavorable climate can limit plant growth and biomass production, thus decreasing process efficiency (Henry, 2000; U.S. EPA, 2000a). The use of green plants as a tool for remediation may increase the availability of contaminants into terrestrial ecosystems. If contaminants are concentrated in above-ground plant parts, these may detach and migrate via wind or water dispersion. Similarly, metal-enriched plants may be ingested by animals, introducing the contaminants to the food chain (Pichtel, 2000).

Harvested plant biomass produced from phytoextraction may be classified by some governments as a hazardous waste, and may therefore be subject to proper handling and disposal requirements.

DEVELOPMENT OF A PRACTICAL PHYTOEXTRACTION TECHNOLOGY FOR CONTAMINATED SOIL

The majority of phytoextraction work is still at the greenhouse level and to a limited degree in the field. In order to progress from the theoretical stage to a commercial phytoextraction technology, the following steps are seen as critical (Chaney, 2000):

Domestication and Breeding of Improved Hyperaccumulator Species

Select hyperaccumulator plant species that have a high likelihood for domestication into a commercial phytoextraction system. Select species that would produce sufficient biomass and accumulate commercially useful levels of hyperaccumulated lead.

Collect Seeds of Plants from the Wild, Bioassay their Ability for Phytoextraction, and Breed Improved Cultivars for a Commercial Phytoextraction System

In order to justify developing improved genotypes for commercial use, a genetic collection is developed.

Identify Soil Management Practices to Improve Phytoextraction Efficiency

Different plants are affected differently by soil pH, fertilization, and other soil management practices. Such practices should be assessed individually for their effect on plant growth and metal uptake.

Develop Crop Management Practices to Improve Phytoextraction Efficiency

All aspects of crop management must be evaluated including:

- Planting practices
- Tillage
- Stand density

- Annual/perennial management
- Weed control
- Harvest schedule and methods
- Seed management

Process the Biomass to Ash in a Form which can be used as Ore or Disposed Safely at Low Cost

Recovery of tissue lead (or other metals) should consider markets for the metal and the feasibility of smelting or other technologies used to recover the metal (Chaney, 2000).

CONCLUSIONS

The success of using green plants to extract lead from contaminated soils requires a thorough assessment of lead levels in soil, an understanding of lead chemistry (i.e., speciation) in the soil, identification of the appropriate plants for lead accumulation, and an appreciation of the potential effects of soil amendments on metal solubilization.

The use of plants capable of absorbing and translocating lead in combination with an appropriate soil amendment may be an effective means of remediating an area having varying lead concentrations. More time and study are required to fully understand the chemical and biochemical mechanisms involved in lead uptake by a plant, and root to shoot translocation within the plant. Likewise, more study is needed to develop and to optimize phytoextraction in field contamination situations.

REFERENCES

Ajwa, H.A., Banuelos, G.S., H.F. Mayland. (1999). Selenium uptake by plants from soils amended with inorganic materials. Journal of Environmental Quality. 27: 1218-1227.

Baker, A.J.M., Brooks, R.R. (1989). Terrestrial higher plants which hyperaccumulate metallic elements – A review of their distribution, ecology and phytochemistry. Biorecovery. 1: 81-126.

Baker, A.J.M., McGrath, S.P., Sidoli, C.M.D., Reeves. R.D. (1994). The possibility of in-situ heavy-metal decontamination of polluted soils using crops of metal-accumulating plants. Resource Conservation Recycling. 11: 41-49.

Banuelos, G.S. (2000). Factors influencing phytoremediation of selenium-laden soils. In: Phytoremediation of contaminated soil and water. pp. 41-59. Eds. N. Terry, G. Banuelos. Lewis Publishing, Boca Raton, FL.

Banuelos, G.S., Ajwa, H.A., Mackey, B., Wu, L., Cook, C., Akohoue, S., Zambruzuski, S. (1997). Evaluation of different plant species used for phytoremediation of high soil selenium. Journal of Environmental Quality. 26: 639-646.

Barber, S.A., Lee, R.B. (1974). The effect of micro-organisms on the absorption of manganese by plants. New Phytologist. 73: 97-106.

Bennett, L.E., Burkhead, J.L., Hale, K.L., Terry, N., Pilon, M., Pilon-Smits, E.A.H. (2003). Analysis of transgenic Indian Mustard plants for phytoremediation of metal-contaminated mine tailings. Journal of Environmental Quality. 32: 432-440.

Berti, W.R., Cunningham, S.D. (2000). Phytostabilization of metals. In: Phytoremediation of Toxic Metals: Using Plants to Clean up the Environment. Eds. I. Raskin, B.D. Ensley. John Wiley and Sons. New York.

Blaylock, M.J. (2000). Field demonstrations of phytoremediation of lead contaminated soils. In: Phytoremediation of contaminated soil and water. pp. 1-12. Eds. N. Terry, G. Banuelos. Lewis Publishing, Boca Raton, FL.

Blaylock, M.J., Elless, M.P., Huang, J.W., Dushenkov, S.M. (1999). Phytoremediation of lead-contaminated soil at a New Jersey brownfield site. Remediation. 9: 93-101.

Blaylock, M.J., Salt, D.E., Dushenkov, S., Zakharova, O., Gussman, C., Kapulnik, Y., Ensley, B.D., Raskin, I. (1997). Enhanced accumulation of Pb in Indian mustard by soil-applied chelating agents. Environmental Science and Technology. 31: 860-865.

Borgmann, U., and W.P. Norwood. 1995. EDTA Toxicity and background concentrations of copper and zinc in Hyalella azteca. Can. J. Fish. Aquat. Sci. 52:875-881.

Bricker, T.J., Pichtel, J., Brown, H.J., Simmons, M. (2001). Phytoextraction of Pb and Cd from a Superfund soil: Effects of amendments and croppings. Journal of Environmental Science and Health. A36: 1597-1610.

Brümmer, G., Gerth, J., Herms, U. (1996). Heavy metal species, mobility and availability in soils. Z. Pflanzenernaehr. Bodenkd. 149: 382-398.

Chaney, R.L., Li, Y.M., Brown, S.L., Homer, F.A., Malik, M., Angle, J.S., Baker, A.J.M., Reeves, R.D., Chin, M. (2000). Improving metal hyperaccumulator wild plants to develop commercial phytoextraction systems: Approaches and progress. In: Phytoremediation of contaminated soil and water. pp. 129-158. Eds. N. Terry, G. Banuelos Lewis Publishing, Boca Raton, FL.

Chaney, R.L., Malik, M., Li, Y.M., Brown, S., Brewer, E.P., Angle, J.S., Baker, A.J.M. (1997). Phytoremediation of Soil Metals. See: http://www.soils.wisc.edu/~barak/temp/opin_fin.htm [6 June, 2000].

Chlopecka, A., Bacon, J.R., Wilson, M.J., Kay, J. (1996). Forms of cadmium, lead, and zinc in contaminated soils from southwest Poland. Journal of Environmental Quality. 25: 69-79.

Ebbs, S.D., Kochian, L.V. (1998). Phytoextraction of zinc by oat (*Avena sativa*), barley (*Hordeum vulgare*) and Indian mustard (*Brassica juncea*). Environmental Science and Technology. 32: 802-806.

Ebbs, S.D., Lasat, M.M., Brady, D.J., Cornish, J., Gordon, R., Kochian, L.V. (1997). Phytoextraction of cadmium and zinc from a contaminated soil. Journal of Environmental Quality. 26: 1424-1430.

Elliott, H.A., Brown, G.A. (1989). Comparative evaluation of NTA and EDTA for extractive decontamination of Pb-polluted soils. Water, Air, and Soil Pollution. 45: 361-369.

Epstein, A.L., Gussman, C.D., Blaylock, M.J., Yermihahu, U., Huang, J.W., Kapulnik, Y., Orser, C.S. (1999). EDTA and Pb-EDTA accumulation in *Brassica juncea* grown in Pb-amended soil. Plant Soil. 208: 87-94.

Evans, L.J. (1989). Chemistry of metal retention in soils. Environmental Science and Technology. 23(9): 1046-1056.

Federal Remediation Technologies Roundtable. (2002). Phytoremediation at the Open Burn and Open Detonating Area, Ensign-Bickford Company, Simsbury, CT. See: http://costperformance.org/profile.cfm?ID=66&CaseID=66

Federal Remediation Technologies Roundtable. (2000). Phytoremediation at Twin Cities Army Ammunition Plant, Minneapolis-St. Paul, Minnesota. EPA 542-R-00-006. June 2000. Abstracts of Remediation Case Studies Volume 4. Washington, D.C.

Glass, D.J. (2000). Economic potential of phytoremediation. In: Phytoremediation of toxic metals: Using plants to clean up the environment. pp. 15-33. John Wiley & Sons, New York.

Grèman, H., Vodnik, D., Velikonja-Bolta, S., Lestan, D. (2003). Ethylenediaminedissuccinate as a new chelate for environmentally safe enhanced lead phytoextraction. Journal of Environmental Quality. 32: 500-506.

Grèman, H., Velikonja-Bolta, S., Vodnik, D., Kox, B., Lestan, D. (2001). EDTA enhanced heavy metal phytoextraction: Metal accumulation, leaching, and toxicity. Plant Soil. 235: 105-114.

Hamon, R.E., Lorenz, S.E., Holm, P.E., Christensen, T.H., McGrath, S.P. (1995). Changes in trace metal species and other components of the rhizosphere during growth of radish. Plant Cell Environ. 18: 749-756.

Heil, D., Hanson, A., Zohrab, S. (1996). The competitive binding of lead by EDTA in soils and implications for heap leaching remediation. Radioactive Waste Management and Environmental Restoration. 20: 111-127.

Henry, J.R. (2000). An Overview of the Phytoremediation of Lead and Mercury. Prepared for U.S. Environmental Protection Agency, Office of Solid Waste and Emergency Response, Technology Innovation Office. Washington, D.C.

Hessling, J.L., Esposito, M.P., Traver, R.P., Snow, R.H. (1990). Results of bench-scale research efforts to wash contaminated soils at battery recycling facilities. In: J.W. Patterson and R. Passion (Eds.) Metal Speciation, Separation, and Recovery, Vol. II. Lewis Publishers, Chelsea, MI.

Huang, J.W., Cunningham, S.D. (1996). Lead phytoextraction: Species variation in lead uptake and translocation. New Phytologist. 134: 75-84.

Huang, J.W., Chen, J., Berti, W.R., Cunningham, S.D. (1997). Phytoremediation of lead-contaminated soils: Role of synthetic chelates in lead phytoextraction. Environmental Science and Technology. 3: 800-805.

Kabata-Pendias, A. (2001). Trace Elements in Soils and Plants. CRC Press. Boca Raton, FL.

Kayser, A., Wenger, K., Keller, A., Attinger, W., Felix, H.R., Gupta, S.K., Schulin, R. (2000). Enhancement of phytoextraction of Zn, Cd, and Cu from calcareous soil: The use of NTA and sulfur amendments. Environmental Science and Technology 34: 1778-1783.

Kerndorff, H., Schnitzer, M. (1980). Sorption of metals on humic acids. Geochemica et Cosmochimica Acta. 44: 1710-1708.

Kumar, P.B.A., Dushenkov, V., Motto, H., Raskin, I. (1995). Phytoextraction: The use of plants to remove heavy metals from soils. Environmental Science and Technology. 29: 1232-1238.

Lasat, M.M. (2000). The Use of Plants for the Removal of Toxic Metals from Contaminated Soil. Prepared for The U.S. Environmental Protection Agency. See: http://clu-in.org/download/remed/lasat.pdf

Ma, L.Q., Rao, G.N. (1997). Chemical fractionation of cadmium, copper, nickel, and zinc in contaminated soils. Journal of Environmental Quality. 26: 259-264.

Means, J.L., Kucak, T., Crerar, D.A. (1980). Relative degradation rates of NTA, EDTA, and DTPA and environmental implications. Environmental Pollution Series B. 1: 45-60.

Nörtemann, B. (1999). Biodegradation of EDTA. Applied Microbiology & Biotechnology 51: 751-759.

Pichtel, J. (2000). Fundamentals of Site Remediation for Metal- and Hydrocarbon-Contaminated Soils. Government Institutes, Inc., Rockville, MD.

Pichtel, J., Anderson, M. (1997). Trace metal bioavailability in municipal solid waste and sewage sludge composts. Bioresource Technology. 60: 223-229.

Pichtel, J., Vine, B., Kuula-Väisänen, P., Niskanen, P. (2001). Lead extraction from soils as affected by Pb chemical and mineral forms. Environmental Engineering Science. 18: 91-98.

Pichtel, J., Kuroiwa, K., Sawyerr, H.T. (2000). Distribution of Pb, Cd and Ba in soils and plants of two contaminated sites. Environmental Pollution. 110: 171-178.

Puschenreiter, M., Tesar, M., Horak, O., Wenzel, W.W. (2001). Rhizosphere manipulation using EDTA to enhance phytoextraction. In: Proc. Int. Conf. on Interactions in the root environment—An Integrated Approach Eds. D.S. Powlsen *et al.*, Rothamsted, UK. 10-12 Apr. 2001.

Ramos, L., Hernandez, L.M., Gonzalez, M.J. (1994). Sequential fractionation of copper, lead, cadmium and zinc in soils from or near Doñana National Park. Journal of Environmental Quality. 23: 50-57.

Reed, B.E., Moore, R.E., Cline, S.R. (1995). Soil flushing of a sandy loam contaminated with Pb(II), $PbSO_4$, $PbCO_3$, or Pb-naphthalene: Column results. Journal of Soil Contamination. 4(3): 243-267.

Reeves, R.D., Brooks, R.R. (1983). Hyperaccumulation of lead and zinc by two metallophytes from mining areas of central Europe. Environmental Pollution. 31: 277-285.

Royer, M.D., Selvakumar, A., Gaire, R. (1992). Control technologies for remediation of contaminated soil and waste deposits at Superfund lead battery recycling sites. Journal of the Air and Waste Management Association. 42: 970-980.

Salt, D.E., Blaylock, M., Kumar, P.B.A.N., Dushenkov, V., Ensley, B.D., Chet, I., Raskin, I. (1995). Phytoremediation: A novel strategy for the removal of toxic materials from the environment using plants. Biotechnology 13: 468-474.

Schmidt, U. (2003). Enhancing phytoextraction: The effect of chemical soil manipulation on mobility, plant accumulation, and leaching of heavy metals. Journal of Environmental Quality. 32: 1939-1954.

Shen, Z.-G., Li, X.-D., Wang, C.-C., Chen, H.-M., Chua, H. (2002). Lead phytoextraction from contaminated soil with high-biomass plant species. Journal of Environmental Quality. 31: 1893-1900.

Smith, L.A., Means, J.L., Chen, A., Alleman, B., Chapman, C.C., Tixier, J.S., Brauning, S.E., Gavaskar, A.R., Royer, M.D. (1995). Remedial Options for Metals-Contaminated Soils. CRC Lewis Publishers. Boca Raton, FL.

Stumm, W., Morgan, J.J. (1996). Aquatic Chemistry. John Wiley & Sons, New York.

U.S. Environmental Protection Agency. (2000a). Introduction to Phytoremediation. EPA 600/R-99/107. Office of Research and Development, Cincinnati, OH.

U.S. Environmental Protection Agency. (2000b). Electrokinetic and Phytoremediation In Situ Treatment of Metal-Contaminated Soil: State-of-the-Practice. Draft for Final Review. EPA/542/R-00/xxx. Office of Solid Waste and Emergency Response Technology Innovation Office, Washington, DC.

Van Benschoten, J.E., Reed, B.E., Matsumoto, M.R., McGarvey, P.J. (1994). Metal removal by soil washing for an iron oxide coated sandy soil. Water Environment Research. 66: 168-174.

Vassil, A.D., Kapulnik, Y., Raskin, I., Salt, D.E. (1998). The role of EDTA in lead transport and accumulation in Indian mustard. Plant Physiology. 117: 447-453.

Yarlaggada, P.S., Matsumoto, M.R., Van Benschoten, J.E., Kathuria, A. (1995). Characteristics of heavy metals in contaminated soils. Journal of Environmental Engineering. 121: 276-286.

Index

Absorption 183
Accumulation 183
Acer rubrum 240
Actinomycetes 97
Activated sludge process 90
Adsorption 137
Advantages 158, 162
Agricultural Chemicals 169
Agrostemma githago 240
Agrostis tenius 180
Alga 196
Alkaline stabilization of sewage
sludge 8
Alliaria officinalis 240
Alternaria triticina 185
Aluminosilicates 64
Ambrosia artemisiifolia 232, 240
Anthropogenic 151
Archaea 116
Aspergillus awamori 186
Asymbiotic nitrogen fixers 97
Avena sativa 232

Bacillus 51
Bacteria 51
Bacterial counts 139
Beta vulgaris 91, 181
Bicarbonate 95
Bioassay 243
Bioaugmentation 160
Bioavailability 26
Biodegradation 159
Biomass-Specific Respiration 199
Biomineralization 69
Bioremediation 44, 157

Biorestoration 160
Biosolids 1, 5
Biosorption 69
Biosphere 27
Biostimulation 159
Biota-to-soil accumulation factor
(BSAF) 30
Brassica 180
—*juncea* 91, 232
—*napus* 232
—*oleracea* 236
Breed Improved Cultivars 243
Breeding 243

Cadmium 35
Cajanus cajan 91
Calcium 95
Carbonates 66
Cation exchange capacity 95
Cellulolytic 97
Chelate 236
Chemical properties of fly ash 173
Chemoautotrophs 50
Chloride 95
Chlorinated Hydrocarbons 164
Chlorinated Solvents 168
Chromium 33
Clostridium 57
Coal 171
Cometabolism 165
Complexation 107
Contamination of soils 44
Correlation 144
Creosote Contaminants 169
Crop Management Practices 243

Crop yield 93
Cropping pattern 91
Crops 89
Crotolaria retusa 186
Crude Oil Spill 168

Daucus carrota 91
Dechloromonas 63
Desulfotomaculum 53
Desulfotomaculum reducens 51
Desulfovibrio 51
Dicoma miccolifera 35
Dioxins 176
Direct dechlorination 165
Disadvantages 163
Domestication 243

Electrical conductivity 95, 178
Electron Acceptor Reactions 164
—Donor Reactions 165
Enterobacter 51
Environmental 148
Excavation 227

Fertilizers 44
Festuca arundinacea 180
Field Study 212
Fluorescein Diacetate Hydrolyzing
    Activity 199
Fly ash aerial deposition 182
Fresh water 89
Freundlich constant 136
Fugitive 225
Fungi 97

Gallionella ferruginea 50
Gasoline Contamination 169
Geissois 35
Geobacter 48
Glomus mosseae 186
Goethite 47
Groundwater 89
—flow 60

Hardwickia binata 179
Heavy Metal Content in Soil 95
Helianthus annuus 180
Helminthosporium oryzae 185
Hematite 47
Henry constant 136
Heterotrophic 50
—aerobic bacteria 97
Hordeum vulgare 232
Horizons 123
How are Biosolids Produced 2
Hyperaccumulator 230

Immobilization 63
Incineration 19, 158
Indirect Transformations 62
Inorganic 4
—Compounds 173
Invertebrates 31
Ion exchange 107
Irrigant 91
Irrigational quality 91
Isotherms 128

Laboratory Study 208
Land application of biosolids 3
Landfill 158
Lead 36
—uptake 230
Lens culinaris 91
Lepidocrocite 47
Leptospermum scoparium 35
Leptothrix 50
Lespedeza cuneata 180
Ligands 58
Limitations 162
Lithoautotrophic 50

Magnesium 95
Medicago lupulina 240
Metal 4, 7
—contamination 9
—Leaching 111

—Mobility 56
—speciation 46
—Tolerance 99
—Toxicity 26
Metalloid 25
—pollution 25
Methods to reduce metals 16
Microbial Biomass 101, 195, 208
—Redox Transformations 46
—Transformations 51
Microbiological Characteristics of Soil 96
Microcosms 69
Microfauna 196
Microorganisms 73, 184
Mineralization 178
Minerals 63
Mining 196
Mobility 36
Municipal solid waste compost 205

Natural Attenuation 161
Nitrate 95
Nitrifying 97
Non-symbiotic Nitrogen Fixation 200

Oil Refinery 90
Organic 4
—Compounds 6, 175
—Matter 107
—Wastes 195, 196

Pathogens 7
Pesticides 44, 169
Phenols 176
Phosphates 68
Phosphorus 95
Physical and chemical properties 5
—properties of fly ash 172
Physiological profiling 141
Phytoextraction 225, 227
Phytoremediation 227
Phytostabilization 227

Pisum sativum 232
Plant Growth 180
Plantago rugelii 240
Podzol 118, 122
Pollutant 4
Pollution 4
Polycyclic aromatic hydrocarbons 175
Potassium 95
Practical uses of Biosolids 2
Precipitation 71
—dissolution 108
Primary substrate 165
Problems of metals in biosolids 9
Protons 109
Protozoa 196
Pseudomonas fluorescens 186
Pseudomonas striata 186

Raphanus sativus 91
Ratio Index Value 213
Reductive dehalogenation 165
Rhizobium 186
Rhizosphere 31
Risk Assessment 105

Saccharum officinarum 91
Sediments 43
Seed Germination 179
Seedling Growth 179
Selenite 54
Sewage Effluent 168
—sludge 2
Shewanella putrefaciens 48
Siderocapsa 50
Smelter 196
Sodium 95
Soil Bacterial Communities 116
—Chemistry 121
—enzyme activities 216
—enzyme activity 101
—fertility 90
—Management Practices 243
—matrix 106

—Respiration 198
Solidification 158, 227
Solubilization 233
Solutions to metal contamination 16
Sorption 135
Sorptive properties 46
Speciation 105
Sphaerotilus 50
Stabilization 227
Storage 19
Streptomyces 56
Stringent control of metal content 17
Sulfides 68
Sulfurospirillum 53
Sulfurospirillum barnesii 53
Surface disposal 19
Sutera folina 35
Synthetic Chelating Agents 234

Taraxacum officinale 240
Temperate Forest Soils 105
Thalspi rotundifolium 232
Thermal Desorption 158
Total dissolved solids 95
Translocation 230
Treatment 19
Trifolium alexandrium 91
Triticum aestivum 91

Vigna radiata 91

Water holding capacity 178
Waterlogging 207

Yield 180

Zea mays 232